RANDOM ACTS

ACTS

of

KINDNESS

by

ANIMALS

RANDOM ACTS of KINDNESS by ANIMALS

Inspiring True Tales of Animal Love

STEPHANIE LALAND

Conari Press
Coral Gables, FL

Cover Design: Megan Werner
Layout & Design: Carmen Fortunato

For permission requests, please contact the publisher at:
Mango Publishing Group
2850 S Douglas Road, 4th Floor
Coral Gables, FL 33134 USA
info@mango.bz

For special orders, quantity sales, course adoptions and corporate sales, please email the publisher at sales@mango.bz. For trade and wholesale sales, please contact Ingram Publisher Services at customer.service@ingramcontent.com or +1.800.509.4887.

Random Acts of Kindness by Animals: Inspiring True Tales of Animal Love

Library of Congress Cataloging-in-Publication Data: 2022935525
ISBN: (print) 978-1-68481-057-4 (ebook) 978-1-68481-058-1
BISAC: PET010000, PETS / Essays & Narratives

Printed in the United States of America

*For Binti-Jua, who showed the world
that animals can be kind.*

*And for my cats, Pie and Little Utz,
whose very existence is an act of kindness.*

Contents

Contents

What separates the beast from us?
Our hands—our brains so ponderous?
What makes us feel superior?
We walk on two legs; they on four.
Abandoned; even pets go feral
Without love, we the world imperil
Through brutish greed, forests we ruin
And make sweet paradise a dune.
As animals respond to kindness
We humans need help with our blindness.
We mustn't be so quick to judge
We, too, must all be tamed by love.

Foreword

As much as I love and cherish our animal friends,
I must tell you it is very difficult for me to read
stories about them or to see them portrayed in
films. Even if the stories have happy endings
and they impart a positive message, I am always
moved to tears.

I've thought about the reasons for all of this
emotion, and I feel it's because I'm so overwhelmed
by the beauty and innocence of these "earthly
angels." What comes to mind when you hear these
words: Compassion, loyalty, devotion, forgiveness,
tenderness, empathy, affection, acceptance,
selflessness, LOVE. These are qualities that our
animal friends have in abundance and oh, how I
wish I could say the same about us humans.

Animals have always touched me in beautiful
and profound ways. My fascination with them
began as a young girl, when I was convalescing
after a horrible automobile accident. My constant
companion was my doggy, "Tiny," and he taught
me how much love, affection, and undemanding
companionship a dog can give. Through the years
I have known and loved so many animals, and
each one has given his or her own unique gift to

me. Literally hundreds of tragic little orphans were rescued by my Pet Foundation and many found their way into my home—just until a proper home could be found, of course. As the song says, "Just one look, that's all it took."

That tender connection—the bond between all living things that we call love—is what this little book is all about. As you read and enjoy these stories, I hope you will recognize all of the wonderful qualities I mentioned and realize that our animal friends are living souls who deserve only the best. It's our duty to see that they get it.

—Doris Day, Spring 1997

Survival of the Kindest

This book is a collection of true stories about animals who have displayed the sort of love and humanity we humans aspire to achieve. Dogs are well known to have put aside their own safety in order to save a fellow animal's life or aid an injured human. But in this book, you will discover the compassion, heroism, and incredible caring, not only of dogs, but also of many other animals, including cats, monkeys, geese, turtles, fish, and even a wild seagull! Some animals have dedicated their whole lives to helping others. In addition to the well-known Seeing Eye and hearing dogs, there are also handi-dogs who help paraplegics.

That animals can act with courage, love, devotion, and compassion is dramatically demonstrated in these stories from around the world. Often animal acts of kindness have a unique poignancy to them—like that of an English cat, who, when barred from the funeral chamber where her beloved mistress was being mourned, laid a dead bird outside the door—the only gift of farewell she knew how to give.

Even if it is not a dramatic rescue, contact with animals can heal the body and nourish the soul. Studies have shown that having a pet enables people to live longer, to recover more frequently from heart attacks, and gives prisoners and juvenile delinquents a way to reconnect with society. From gorillas who speak to us through sign language to pets who communicate so much love throughout a lifetime, without ever saying a word, animals bring joy to our lives and comfort to our spirits.

Four years of research have gone into this book. These stories have made me laugh, made me cry, amazed and surprised me. Finding each one was like discovering a delicate little treasure more precious than gold or jewels, because each spoke to me of the virtues common to all life. When we discover a little more goodness in others, we somehow discover more of it in ourselves as well.

Never again will I settle for the phrase "only an animal" to describe one of my fellow creatures. The animals, themselves, through their loyalty, bravery, commitment, and love, have stated, more eloquently than any words might say, that we are all sentient and conscious beings, each holding within us that little spark of the divine.

By now everyone has heard of the incident that took place at a zoo near Chicago, an unexpected rescue that captured the attention of the whole world and caused people to rethink the way they looked at animals. During the summer of 1996, spectators at the Brookfield Zoo were shocked to witness a three-year-old boy fall twenty-four feet into the gorilla compound, striking his head on the concrete. The boy lay unconscious, at the mercy of the seven huge gorillas in the compound.

As the crowd watched anxiously, one of the gorillas approached the boy. It was Binti-Jua, a mother gorilla, carrying her own infant on her back. The helpless spectators tensed as the gorilla came closer and closer to the unconscious boy. They were too far away to separate them if she chose to do him harm. As Binti-Jua picked up the boy in her hairy arms, the boy's mother screamed out, "A gorilla's got my baby!"

Rather than harming him, the gorilla cradled him in her arms. Keeping the other gorillas away, she carefully carried him to a door at the side of the compound which she had seen humans use many times and where the zookeeper could reclaim him.

Binti-Jua (Swahili for "daughter of sunshine") became a national hero, thanks to a videotape made by visitor Bill Lambert that captured the entire event. Audiences around the world watched a zoo animal rescue a human boy with obvious tenderness. Zoo attendance rose dramatically and gifts of money and bananas flooded in from around the world.

Critics who said that Binti-Jua was merely acting from maternal instinct (as if all mothers don't) were silenced when a videotape was broadcast of a similar incident that happened ten years before. On that occasion, a child who had fallen into a gorilla compound was approached by a male gorilla. Instead of hurting the child, the very large and potentially dangerous male gorilla could be seen gently stroking the fallen child's head.

Why should we be so surprised? Anyone lucky enough to be raised with a dog or cat knows that animals are fellow beings with personalities and a tremendous capacity for love and devotion. You can look into their eyes and sense another soul in there looking back at you. We all love to share stories of our own animals' antics or heroism.

The fact that wild animals are also capable of compassion and understanding should not be too difficult to accept. Yet, when we witness wild animals demonstrating remarkable acts of kindness and courage, we are taken aback. Such cases make us uneasy, impel us to reach, like Binti-Jua's critics, for rationalizations that let us believe animals are mere soulless automatons.

When animals display extraordinary humanity, it is hard for us to continue to treat them "like animals." And yet, in the animal kingdom there are many examples of compassion taking precedence over "survival of the fittest."

Indeed, the naturalist Petr Kropotkin, a Russian prince and contemporary of Charles Darwin, noticed such behavior and developed a very different theory from Darwin. Darwin purported the theory of "survival of the fittest," a dispassionate theory that describes life on earth as a desperate contest to kill or be killed—to eat or be eaten. In contrast, Kropotkin's theory championed "survival of the most cooperative."

Why such different views of animals? Darwin developed his theories predominantly on an island where the air was warm, food was abundant, and animals were at risk from overpopulation. Petr Kropotkin, however, went north to Russian Siberia to learn about animal behavior. There, he realized that when the environment is more adversarial, animals must learn to work together for survival. Animals such as wolves, which have had to face biting cold and snow, learn to hunt together and build supportive social structures. Kropotkin's research indicated that cooperation, communal living, and mutual assistance are just as important to life on earth as sharper teeth, longer claws, or bigger muscles.

Even in warmer climates there are many examples of animal cooperation if you look for them, such as beavers working together to build a dam to stop a river's flow. Since we know that what we look for affects what we see, we must guard against the prejudicial veils of theory and look afresh at the animal kingdom.

That's the purpose of this book—to open our eyes to the compassionate and cooperative nature of the animals with whom we share this earth. It includes stories not only of love and heroism, but also of works of great beauty that animals have created. And sprinkled throughout the book are suggested acts of kindness we can offer to the animals in our lives in appreciation of all of their devotion and care.

This book began to take shape over ten years ago, although I didn't know it at the time. A woman called me on the telephone to tell me about an incident that took place on her childhood farm many years before. She had gotten my name from an article I had written about animals and had taken the time to call me up. As she shared her story of love and compassion, I was overwhelmed by the sweetness, loyalty, and self-sacrificing love that passed between two dogs—one her's, the other her neighbor's. At the end of the conversation, we were both crying. She thanked me for listening and then added, "I've waited thirty years to tell someone this story." And so I have included the story of Brownie and Spotty here, the first of a series of stories that have gently refocused my life.

—Stephanie Laland, Spring 1997

Unspoken
Love

*"Not to hurt our humble brethren is our
first duty to them, but to stop there is not
enough. We have a higher mission—to be of
service to them wherever they require it."*

—*Saint Francis of Assisi*

B rownie and Spotty were neighbor dogs who met every day to play together. Like pairs of dogs you can find in most any neighborhood, these two loved each other and played together so often that they had worn a path through the grass of the field between their respective houses.

One evening, Brownie's family noticed that Brownie hadn't returned home. At first, they weren't too concerned because he had disappeared before. Assuming he was just out roaming, they didn't look for him. But Brownie didn't show up the next day and by the next week he was still missing.

Curiously, Spotty showed up at Brownie's house alone, barking, whining, and generally pestering Brownie's human family. Busy with their own lives, they just ignored the nervous little neighbor dog. Finally, one morning Spotty refused to take "no" for an answer. Ted, the father of the family with whom Brownie lived, was steadily harassed by the furious, adamant little dog. Spotty followed him about, barking insistently, then darting toward the empty lot and back as if to say, "Follow me! It's urgent!"

Ted followed the frantic Spotty across the empty lot as Spotty paused to race back and bark encouragingly. The little dog led the man under a fence, past clumps of trees, to a desolate spot a half mile from the house. There Ted found his beloved Brownie alive, one of his hind legs crushed in a steel leghold trap. Horrified, Ted now wished he'd taken Spotty's earlier appeals seriously.

Then Ted noticed something quite remarkable.

Spotty had done more than simply lead Brownie's human to his trapped friend. In a circle around the injured dog, Ted found an array of bones and table scraps—which were later identified as the remains of every meal Spotty had been fed that week!

Spotty had been visiting Brownie regularly, in a single-minded quest to keep his friend alive by sacrificing his own comfort. Spotty had evidently stayed with Brownie to protect him from predators, snuggling with him at night to keep him warm and nuzzling him to keep his spirits up.

Brownie's leg was treated by a veterinarian and he recovered. For many years thereafter the two families watched the faithful friends frolicking and chasing each other down that well-worn path between their houses.

Scarlett, an ordinary-looking short-haired cat, was not really noticed at first in the confusion that surrounded the burning building where she had been living with her small kittens. But Scarlett proved to be a devoted mother and a hero.

Overcoming every animal's innate fear of fire, she forced herself to go back into the roaring flames and billowing poisonous smoke of the building and retrieve each of her precious kittens. Five times Scarlett returned to the ferocious heart of the blaze to get each of her babies out. Scarlett's fur was singed off and her eyes seared shut by the flames, and yet she somehow managed to carry each tiny kitten to safety across the street. There she was observed taking a head count by touch because she could no longer see them.

Firemen finally found her and realized what had happened. Much of her body had been burned in the course of getting her kittens out. She was taken to the local animal shelter and separated from her kittens because she could not feed them due to her burns. After a local TV station featured her tale of heroism, the shelter received over 10,000 calls from people wanting to adopt her. A week later, she was reintroduced to her kittens and the joyful reunion was broadcast across the nation. Scarlett had her kittens back at last and she licked each youngster in turn, purring happily. One of her fireman rescuers who had dropped by to visit said, "Just to see her do that same head count almost made me cry."

      .ʻ.   🐾   .ʻ.

Because mother dolphins nurse their young for so long—eighteen months or about as long as human mothers nurse—the mother-child bond is very strong. Many times a dolphin will not desert another dolphin who is in trouble even if it costs them their own life. When infant dolphins are trapped in tuna nets, their mothers will try desperately to join them. Then the mothers will cuddle close to their babies and sing to them as they both drown. The tuna industry's official acknowledgment of this remarkable phenomenon is that most of the dolphins killed are mothers and infants. Although they are an improvement, even the "dolphin-safe" nets used to catch tuna continue to kill dolphins.

A man raised a pet deer that, when young, was so tame it even liked to ride in the car with him like a dog. During the hunting season they would pass other cars full of hunters whose startled gazes the deer would placidly return.

The same deer, when it was a bit older, once came upon a lost deer hunter in the woods. The hunter, completely disoriented, was startled by the deer's friendliness—it trotted right up to him, obviously tame. Figuring he had nothing to lose, the hunter decided to follow the deer. Sure enough, the deer led the man back to his house where someone opened the door. The deer casually walked into the house.

The deer's human guardian gave directions to the now thoroughly abashed hunter, while the deer that had guided him to safety fell asleep on the couch.

"If you pick up a starving dog and make him prosperous, he will not bite you. This is the principal difference between dog and man."

    —Mark Twain

A twelve-year-old boy, Rheal Guindon, had gone on a fishing trip with his parents to Ontario, Canada, where they enjoyed camping out. The lake was in a remote area and there was no one else around. The nearest town was miles away.

One day, Rheal stayed on shore while his parents took a boat out on the lake to catch some fish. As he watched from the shore, his parents' boat suddenly overturned. They struggled in the water and before their son's eyes began to drown. He had no idea how to help them and could only watch helplessly, shouting desperately from the shore. Then their frantic cries ceased and all was silent.

Grieving, the boy numbly tried to walk to the safety of a distant town but the sun was setting. Terrified, he realized he had to face a night alone in the woods. As night fell, he lay down on the freezing earth, weeping and shivering.

Suddenly, as if an angel had heard his cries, he felt a furry body press against him. Rheal couldn't tell what it was, perhaps a dog, but just being next to something warm and breathing helped to ease his pain. He put his arm around the animal and, consoling himself in its warmth and closeness, fell asleep.

In the morning, he woke up to find three wild beavers huddled against him and across his body. They had kept him from freezing to death

during the night when the temperature had
fallen below zero.

∴ 🐾 ∴

*Saint Bernards have been performing rescue work for at
least three centuries and have saved thousands of lives.
They have wide, almost weblike toes that enable them to
walk on snowdrifts up to sixty feet deep. Saint Bernard
dogs often travel in packs when doing their rescue work.
When they find a fallen traveler, two dogs will lie down,
one on either side of the traveler's body to warm him
while a third licks his face to awaken him. A fourth dog
goes for help to guide a rescue party to the location.*

One day a half-starved puppy wandered through the gates of the maximum security prison Sing Sing. He was so undernourished that the fur hung off his body; he looked as if he were wearing poor and ill-fitting clothing. Named "Rags" because of his appearance, he was an instant hit with the inmates, who saved food from their meager meals for him. The men of Sing Sing often had no friends or family who cared to write to them, and they felt abandoned and alone. Rags became a true friend to many, and he loved all of the prisoners. And yet, Rags was aloof to the warden and his family, and he growled at all the guards. He would exercise with the prisoners, and when they had softball games, Rags would bark madly and run around the field with glee.

Every night Rags would leave the prison and return in the morning. Every night but one. This time Rags followed a prisoner to his cell and kept vigil there all night. The next morning, the prisoner confessed to his fellow inmates; "That dog just saved my life." For when his parole had been denied, the man had decided to end his life. Yet every time he's move to wind the bedsheets to hang himself, Rags would softly growl outside his cell. Knowing that if he continued, Rags would bark and alert the guards, the prisoner was unable to act on his plan. At last he realized that there was someone

who really cared if he lived or died—Rags. Secure
in this knowledge, he had gratefully chosen to live.

*"I refuse to eat animals because I cannot nourish
myself by the sufferings and by the death of other
creatures. I refuse to do so, because I suffered so
painfully myself that I can feel the pains of others
by recalling my own sufferings."*

—*Edgar Kupfer, German imprisoned by the
Nazis for his beliefs, writing secretly from his
hospital bed in Dachau*

Toto was a tame chimpanzee and longtime companion of Mr. Cherry Kearton. When Cherry fell desperately ill with malaria, Toto sat up with him day and night. As Cherry grew weaker, Toto learned to bring a glass of quinine, the medicine needed to control the disease, to his friend.

While he was recovering but before he could rise from his bed, Cherry would signal Toto that he wanted to read. Toto learned to put a finger on each book on the shelf until the man said "Yes." Then the chimpanzee would pull the indicated book out of the shelf and carry it over to his patient. Sometimes, when Cherry fell asleep with his boots on, Toto removed them for him.

In 1925, Cherry wrote: "It may be that some who read this book will say that friendship between a man and an ape is absurd, and that Toto being 'only an animal' cannot really have felt the feelings that I attribute to him. They would not say it if they had felt his tenderness and seen his care as I felt and saw it at that time. He was entirely lovable."

Scientists have found that rhesus monkeys refuse to pull the levers that deliver their food pellets when they see that pulling the lever also delivers an electric shock to a fellow monkey.

A canary and a cat grew up together and became close friends. They would play together and when the cat slept, the canary perched on his belly. None of the typical cat-bird animosity existed between them.

Joan, their owner, came home one day to find her canary dead on the floor. Convinced that her cat had finally succumbed to instinct and killed her canary, Joan screamed furiously at the cat sitting nearby and tried to swat her but the cat dashed out the door.

Later, on examining the bird, Joan realized that it had simply died of old age; there were no teeth marks, no sign of attack whatsoever. Guiltily, she called for her cat but the falsely accused animal would not return.

The cat's habit was to come home every evening by eight o'clock, but this time she did not appear. As the hours passed, the woman grew more and more concerned.

Finally at midnight, to her great relief, she heard a scratching at the door. When she opened the door, there was the cat on the threshold, delicately holding a live fledgling in her mouth. Gently the cat placed the little bird on the floor at the woman's feet, backed away, and sat down to watch her human expectantly.

The young bird blinked and peeped. The cat had obviously stolen the fledgling from its nest. The cat looked hopefully up at Joan to see if the new bird would ease her sorrow. The cat's look seemed to say, "Can we be friends again now? I've brought you another bird."

Long ocean voyages were once very difficult and one reason was because mice and rats would eat or foul all the stored food supplies. Ships had to travel from port to port to restock, clinging close to land. Then, as cats occasionally crept aboard, sailors discovered that not only did they make good mates (because their lithe bodies could roll with the ship's motions) but the mice and rat population was greatly decreased. It became known bad luck to chase a cat off a ship that it chose to board. Thus, the tradition of "ship's cat" was born.

An English trapper came to America long ago and fell in love with the country and with a lovely Iroquois woman named Anahareo. One day he found a mother beaver in one of his traps and nearby two tiny beaver kits. At his wife's urging, he took the two tiny beaver babies home with him. During the course of raising them he realized he would never hunt animals again. At the time of this decision he wrote: "Their almost childlike intimacies and murmurings of affection, their rollicking good fellowship not only with each other but ourselves, their keen awareness, their air of knowing what it was all about. They seemed like little folk from some other planet, whose language we could not quite understand. To kill such creatures seemed monstrous. I would do no more of it."

> "Thus godlike sympathy grows and thrives and spreads far beyond the teachings of churches and schools, where too often the mean, blinding, loveless doctrine is taught that animals have no rights that we are bound to respect, and were made only for man, to be petted, spoiled, slaughtered or enslaved."
>
> —John Muir

"Go away King," Pearl Carlson said sleepily as her German shepherd dog pulled at her bedding and attempted to rouse her. "Not now, I'm trying to get some rest." Pearl vaguely wondered what King was doing in her bedroom at three o'clock in the morning, since he was usually locked in another part of the house at night. It was Christmas night and the sixteen-year-old girl had been looking forward to a good night's sleep after an exciting day.

Pearl sat up in bed to give the barking dog a good push and realized that smoke was filling her room—the house was on fire. Bolting out of bed, she ran in panic to her parents' bedroom and awoke them both.

Her mother, Fern, got up at once and told Pearl to escape through her own bedroom window while she helped her husband, Howard, out of their window. But Howard Carlson had a lung condition and could not move quickly. Pearl headed back to her own bedroom but somehow wound up in the living room where the fire was at its worst.

"I'm going after her," Howard said, but his wife, knowing his lung ailment made this impossible, told him to escape through the window while she went for Pearl. Fern ran blindly toward the smoke-blackened room. Pearl was standing there, frozen in confusion. Fern led her to safety but then realized that neither Howard nor King had gotten out of the house. Fern ran back into their bedroom and found Howard collapsed on the floor with King by his side. Fern and King struggled to lift Howard and finally the two of them managed to get the nearly unconscious man to safety. Fern later said she could not have moved Howard without King's help.

King and his family were saved. King had badly burned paws and a gash on his back, but seemed otherwise healthy. Yet the day after the fire, King would not eat his dog food.

The neighbors had come by with sandwiches and refreshments and were helping to rebuild what they could of the house. Then King did something he had never done before: He stole one of the soft sandwiches. Something was wrong. The Carlsons looked in King's mouth and saw that his gums were pierced with painful, sharp wooden splinters. That terrible night, King had, with sheer desperate force, chewed and clawed his way through the closed plywood door that separated him from his family. A door had been left open for King to the outdoors so he could easily have just saved himself. Instead, he chose not to flee but to gnaw and smash through the door to face the blinding fire and choking smoke to rescue his friends. Now the family knew how King had gotten into the house.

The splinters were removed and King recovered fully, although his pads had been so burned that even a year later it was painful for him to walk on a hot sidewalk. I asked Fern Carlson later, as she recounted the incredible tale of King's bravery, if there were any changes in the way King was treated after the fire. "Oh yes," she said. "The neighbors all fed King steak and roasts until he got really fat!"

"Where is man without the beasts? If all the beasts were gone man would die from a great loneliness of spirit."
—Chief Seattle

Observers have repeatedly noticed that animals in the wild do not live solely by "tooth and claw" but regularly show compassion for their fellows. For example, a newborn elephant is raised by both its mother and a special "auntie." This second mother acts as helper, babysitter, and guardian. This relationship occurs naturally in the wild and helps to protect the helpless youngster from tigers.

Once, when an old bull elephant lay dying, human observers noted that his entire family tried everything to help him to his feet again. First, they tried to work their trunks and tusks underneath him. Then they pulled the old fellow up so strenuously that some broke their tusks in the process. Their concern for their old friend was greater than their concern for themselves.

Elephants have also been observed coming to the aid of a comrade shot by a hunter, despite their fear of gunshots. The other elephants work in concert to raise their wounded companion to walk again. They do this by pressing on either side of the injured elephant and walking, trying to carry their friend between their gigantic bodies. Elephants have also been seen sticking grass in the mouths of their injured friends in an attempt to feed them, to give them strength.

E ven a duck can be a hero. On November 27, 1944, the Allies launched an air attack against Freiburg, Germany. Unfortunately, the town's air-raid sirens weren't working.

The local inhabitants would not have had a chance for survival were it not for a vocal duck who lived in Freiburg's main park. The residents had noticed that just before an air raid, animals would sometimes begin vocalizing hysterically, as if they somehow sensed the distant bombers long before the warning system. On this occasion, although the sirens failed, the duck's frenzied squawking drove many hundreds of people into the air-raid shelters.

Unfortunately, the duck was killed in the bombing, but after the war, when Freiburg was rebuilt, the survivors commemorated their web-footed savior with a monument in the new park.

Many animals—from robins and thrushes to vervet monkeys—utter a piercing warning cry when a predator approaches. The shriek enables others of its kind to hide or flee, even though it also attracts the predator's attention, sometimes resulting in the sentinel's death.

W hile researching animal behavior for her book *Mongoose Watch*, British ethologist Anne Rasa was surprised to discover that when a dwarf mongoose became ill with chronic kidney disease, he was treated differently by his peers.

The other mongooses permitted the ill animal to eat much earlier than he normally would have, considering his rank in the mongoose social order. To Rasa's astonishment, the sick mongoose was even allowed to nibble on the same piece of food that the dominant male was eating—something that would never occur normally.

When the ill mongoose lost his ability to climb, the entire group of mongooses gave up their decided preference for sleeping on elevated objects such as boxes. Instead, they all opted to sleep on the floor with their sick friend.

⁘ ❀ ⁙

A mongoose had injured its front paw so that it could no longer capture food. While they did not overtly bring food to her, the other mongooses, upon seeing her plight, started to forage for food near her. They did not offer her food, as this was against mongoose etiquette, but made sure they were close to her so when she asked they could share.

During the Civil War, an eagle, stolen from his nest while still a baby, was sold to a man who joined the Union army and went to fight against the Confederates. Growing up with the army, the young eagle soon became their mascot. Possibly because of his beaky profile, the troops named him "Old Abe" after their hero Abraham Lincoln.

Old Abe soon proved his loyalty. Once, as his platoon was about to enter a wood, the eagle began swooping over their heads so crazily that their commander called a halt. When they tried to resume their advance, he began flying into the faces of the men on the front line. Spooked, the soldiers decided to fire a few rounds into the woods. Instantly, Rebel troops, lying in wait, fired back and the Union troops took cover. Were it not for Old Abe's warning, the Union soldiers would have walked into a trap and been massacred.

Thereafter, whenever his regiment fought, Old Abe appeared and his troops were always victorious. The eagle became such a symbol that Confederate General Major Sterling Price once remarked that he would "rather kill or capture the eagle than take a whole brigade."

Finally old Abe was wounded and the "Yankee Buzzard," as the Confederates called him, became part of the North's public relations campaign.

Old Abe traveled around the United States and was a featured attraction at parades and patriotic events. When his local regiment returned home to Wisconsin after the war, a banquet was held in the State Capitol building with Old Abe as the guest of honor.

For the last fifteen years of his life, Old Abe lived in a cage in the Capitol Building of Madison, Wisconsin. One day, a fire broke out in the building and he roused the watchman by banging his tin cup against his cage. Again Abe had been a hero. Although his warning was enough to save the building, the watchman forgot the bird in the excitement of the fire and Old Abe died of smoke inhalation.

"I am in favor of animal rights as well as human rights. That is the way of the whole human being."
—Abraham Lincoln

"I wish people would realize that animals are totally dependent on us, helpless, like children; a trust that is put upon us."
—James Herriot

The dogs of the neighborhood were apparently not on guard duty this particular day. Indeed, the one animal that was making a fuss about whatever was going on across the street was a cat named Emily.

She paced back and forth in the front room window, growling ominously to signal that something, in her opinion, was very wrong. Drawn by her vocalizations, Emily's family looked across the city street just in time to see a burglar climbing in a neighbor's window. Emily's family alerted the police, the man was apprehended, and all was well.

When he dined, Winston Churchill would have his servants bring his cat Jock to the table to share dinner. Churchill considered Jock one of his more agreeable dining companions.

While collecting specimens of birds, a naturalist named George Romanes shot a tern, which fell into the sea. At once, other terns gathered around the fallen bird, "manifesting much apparent solicitude, as terns and gulls always do under such circumstances."

The wounded bird began drifting toward the shore accompanied by his companions. Edward walked towards the downed bird to collect it. To his utter shock, two of the attendant terns grasped the wounded bird, one tern holding each wing, and lifted him out of the water. The two terns began to carry the injured bird towards the sea. They carried him about seven yards and carefully set him down again, where he was then taken up again by a fresh pair of birds and carried a little farther. In this way, the terns continued to carry the injured bird alternately, until they had brought him to a sea rock at some distance from the human attacker.

Shaking his head, the man started toward the bird again with the intention of capturing it. To his surprise, a great swarm of birds descended in front of him, obstructing his path. As he pushed through the birds, getting closer to the rock, he watched as two birds again took hold of the wounded bird's wings and carried him out to sea, far beyond the man's reach.

"This, had I been so inclined," Romanes wrote, "I could no doubt have prevented. Under the circumstances, however, my feelings would not permit me; and I willingly allowed them to perform an act of mercy which man himself need not be ashamed to imitate."

> "The animal shall not be measured by man. In a world older and more complete than ours, they move finished and complete, gifted with extensions of the senses we have lost or never attained, living by voices we shall never hear. They are not brethren, they are not underlings; they are other nations, caught with ourselves in the net of life and time, fellow prisoners of the splendor and travail of the earth."
>
> —Henry Beston

Ways to Return the Kindness

❧ If you have an animal that plays in your backyard, try planting the herbs pennyroyal, rosemary, fennel and rue along with eucalyptus trees to act as a natural flea repellent. Not only will it help keep your backyard free of fleas, but when your animal sleeps or lies in these herbs, the oils will discourage any fleas from staying in their fur.

❧ Cap your chimney so small animals do not fall in.

❧ If you live near a dog that barks a lot, it might just be lonely. Some people acquire dogs and then do not pay proper attention to them. Speaking politely to the person in charge, ask if you can walk the dog occasionally. Then introduce yourself to the dog and walk it and play with it as often as you can. Remember to keep your commitment once you start, as you will become a center of the dog's life. Of course, real neglect is against the law and should be reported to the SPCA (Society for the Prevention of Cruelty to Animals).

❧ Birds frequently fly into large picture windows and injure themselves. You can help prevent this by placing cut-out silhouettes of hawks in your window. The birds know instinctively to fear the hawk shape and won't fly near it. Streamers and chimes can also be used.

❧ If you find a baby bird on the ground, do not immediately assume it has fallen out of its nest. If it has feathers it is a fledgling and this may just be a normal part of its learning to fly. If there are no cats or other predators around, one thing you can do is place nesting materials in a box, nail it securely to a tree and place the little bird in it. Watch to see if the mother bird comes for it and make certain it is not in danger. Under no circumstances should you toss it into the air to "help" it fly. Call your local Native Animal Rescue for advice if you need it. If they're not in the phone book, you can get their number from your local SPCA. Mealworms or baby food run through a blender are usually safe bets for feeding.

At the end of every circus performance at the Wirth circus in Australia, all the animals were marched around the ring for the Parade of Animals. Elephants, big cats, playful seals, and other large animals strutted around the circus floor. It was a majestic moment, most impressive to everyone—only this day something went wrong.

A little girl, drawn by the color and excitement of the procession, had wandered out of the stands and down to the ring. Her mother screamed in horror as she saw her child, who had been sitting next to her, approaching the huge animals where she would surely be trampled to death or, perhaps, eaten before her very eyes.

As the crowd held its breath, Alice, the circus's most famous elephant, stepped forward and gently picked the child up with her trunk. Gracefully, as she had held so many circus performers before, Alice delivered the child safely back to her mother's arms.

Alice was a very intelligent elephant who became legendary for her long lifetime of heroism and remarkable adventures. Born about 1850, Alice was raised in Burma where she hauled logs, until she was sold to the Wirth circus. Once, while Alice was moving circus wagons (elephants were always used to carry the really heavy equipment), she saw a team of bulls trying to pull a wagon stuck on train tracks. An express train was due in a few minutes; if the wagon were still on the tracks, a tragedy would occur. But no matter how hard the bulls pulled, their strength was not enough to dislodge the wheels from the tracks. At once, without any instructions from her handlers, Alice put down her load, ran to the wagon and pushed her body against it, easing it over the tracks. Minutes later the train roared by.

Alice lived to be over one hundred years old. As she grew older, her health failed and for awhile she was boarded at the Melbourne Zoo. But solitude did not agree with her and she worsened. She missed her human friends and the other circus elephants and started wasting away. The zoo, which her handlers had meant to be a rest home, was only a prison for her. She was dying, not just of old age, but of loneliness, too. Finally, a trailer truck was sent to the zoo to bring her back. Alice sensed she would soon see her friends, but even before the other elephants could see her, they picked up her scent and excitedly trumpeted their welcome. Alice trumpeted back, nearly wild with anticipation of seeing her friends again.

Back in the herd at last, Alice immediately responded by fondly caressing the others with her trunk. Her listlessness fell away as if by magic; a sprightliness returned to her step. However, Alice never worked or performed again; her health would not permit it. For three years, until April of 1956 when she died, Alice traveled as "guest of honor" with the circus.

∴ ❀ ∴

Racehorses often form bonds with animals of other species, such as a goat or a cat with whom they share a stable, and will not run well if they are deprived of their friends' company.

Washoe, a female chimpanzee, was the first of her species to be taught sign language. One day, a new chimpanzee was introduced to Washoe's compound. The newcomer panicked, jumped over an electric fence, and landed in the moat surrounding the compound. To the keeper's surprise, Washoe followed, also braving the electric fence, and landed on a strip of land alongside the moat. Reaching into the moat, she pulled the frightened newcomer to safety.

The kind chimpanzee would have made a good mother, but unfortunately, it was not to be, for Washoe's attempts at motherhood ended in tragedy. She had a baby that died four hours after birth because of a defective heart. Three years later, she had another baby that died of pneumonia after two months. Washoe became depressed.

The researchers working with the unhappy Washoe decided to try to find a baby chimp for her to adopt. They located a ten-month-old chimpanzee and had it shipped to them for Washoe to mother. When the researcher signed, "I have a baby for you," Washoe displayed great excitement. She stood, puffing up her fur, hooting, and displaying the sign for "baby" repeatedly.

When she finally met the youngster, she was miserable. It was obvious she had been hoping that the humans had somehow revived her own baby. Her fur flattened back down and she refused to pick the baby up. She showed no interest in the newcomer and would only lethargically sign "baby." After an hour, her heart began to open up to the little one and she tried to get him to play with her. By nightfall, she was coaxing him to sleep in her arms. At first he refused, but by the next morning they were seen cuddled up together.

Eventually she and the other chimpanzees in their group taught him fifty hand positions of sign language without any help from humans.

⠒ 🐾 ⠒

Washoe's own made-up word for watermelon was "drink-fruit."

Chimpanzees have been observed to sit and watch particularly beautiful sunsets together. In one case, they were even seen to be holding hands.

A thief thought he was very clever to break into the circus office and steal the circus's money. But as he left the office with his bag of loot in hand, he was shocked as an elephant's trunk snaked out of nowhere, tightened around his torso, and hoisted him into the air. The thief was held in the air by the vigilant elephant until the police arrived and the money was handed over. The next day newspapers carried pictures of the sheriff pretending to pin his own badge on the crime-stopping elephant.

"In my humble opinion the intelligence of the elephant...is superior to all land mammals, except, perhaps the human animal, and sometimes I have my doubts about humans."
—George Lewis

Bibs the canary lived with an elderly lady who had a niece who lived next door and checked on her each night to make sure she was all right. A warm and sweet friendship had blossomed between the old woman and the tiny bird. At breakfast each morning, they shared toast and Bibs liked to sip whatever beverage the woman was having.

One rainy night, seeing that her aunt's lights were on and assuming everything was fine, the niece retired with her husband for the night rather than going over to the aunt's house. As the couple relaxed cozily by a fire, they were startled by an odd tapping at the window.

At first they assumed it was a windblown branch, but the tapping grew louder and continued persistently, followed by a strange cry. Finally, the niece went to the window, pulled back the curtains and found Bibs, who had been furiously beating on the window and chirping.

The tiny yellow bird had escaped from the aunt's house and flown through the storm to the next house. There it had pecked at the window with such desperate fury that it collapsed in exhaustion and died before their eyes. Now completely alarmed, the niece and her husband raced over to the aunt's house.

They found the old lady lying unconscious on the floor in a pool of blood. She had slipped and fallen, striking her head on a table corner. Her niece rushed her to the hospital.

Because of her little bird's loyalty and determination to get help, even at the sacrifice of his own life, the woman's life was saved.

∴ ❀ ∴

Gratitude is an emotion often assumed to belong exclusively to humans. However it is so common in parrots that cunning parrot-traders often create a dangerous or frightening situation for parrots just so they can rescue them from it. The parrots then mistakenly give their love and trust to the dealers and are easier to control.

In May 1977, a family in Malmö, Sweden, learned just how far out on a limb a dog will go for a human. The father, Leif Rongemo, returning from the kitchen to the living room of his third floor apartment, discovered that the casement window was open and his two-year-old daughter was missing from the living room.

When he looked outside, he saw the street thirty-six feet below—and his little daughter crawling on all fours along a narrow concrete ledge that circled the building. He stifled a cry of alarm to avoid startling her, because an extraordinary kind of rescue was already under way.

Following just behind the baby girl was the family's Alsatian dog, whining softly with distress. Both dog and daughter were far beyond reach and there was no room on the narrow ledge for either to turn around and crawl back to the window. The father quickly realized that if he attempted the ledge, all three would probably die. He called his wife to help from the window while he rushed down to the street to try to catch the child.

As the baby girl crawled farther away from the window and safety, the dog pushed forward determinedly and at last seized the child's diaper in his jaws. To the amazement of those who had gathered on the street below—they were attempting to create a makeshift net to catch the child—the dog then shuffled carefully backwards, inch by inch, pulling the little girl back toward the window.

The heart-pounding backwards journey took three minutes, until the mother could snatch her child. The dog then leapt into the room, proudly wagging his tail.

The family had been thinking of getting rid of the dog because they were concerned he might be too big to keep around a small child. His bold rescue of their daughter, however, made him a most treasured member of the family.

"You think that these dogs will not be in Heaven! I tell you they will be there long before any of us."

—Robert Louis Stevenson

In India, monkeys, though mischievous, are revered as descendants of Hanuman, the monkey servant of God. In many places they are allowed to roam freely among humans, whom they seem to regard simply as another troop of monkeys. Monkeys regularly visit Hindu Temple of Hanuman for handouts or work in small groups for human trainers, who, in return, sometimes teach them the Hindu philosophy of the soul in stories and pictures of the noble Hanuman.

One afternoon, near New Delhi, a mother was walking with her infant in her arms along the shore of the Payaswami River. The infant slipped from her grasp, and to her horror, fell into the flowing water. The mother screamed in anguish; she did not know how to swim.

A group of monkeys were watching from a nearby tree. Suddenly, one monkey leaped from the tree directly into the river and swam toward the baby. Braving the current, the little monkey dragged the human infant to the shore where the frightened mother waited.

The monkey placed the child at its mother's feet and scrambled back up into the tree.

A New Yorker named Eldon Bisbee lost his French poodle during a severe snowstorm. After searching through the storm for hours, Eldon returned home in sorrow. Unable to sleep, he pondered his loss.

At 3 a.m. Eldon's doorbell rang. His heart leapt.

It was a cab driver asking, "Do you have a poodle?" The driver had the missing poodle in his cab and had obtained the man's address from her collar. While Eldon was reunited with his poodle who joyfully licked him, the cabby told of the amazing rescue.

The cab driver had been cruising through the storm when a German shepherd jumped in front of the cab, planting himself before the headlights and refusing to let the cab pass. The driver blew the horn threateningly, but the dog wouldn't move.

When the cab driver rolled down his window to scold the dog, the German shepherd approached and whined expressively. Still whining, the dog ran a few feet off in the snow, then hesitated, looking back so pleadingly that the cabby, though a stranger to the dog, followed. The dog led the cabby to the half-frozen poodle who lay immobilized in a snow bank. She had been injured by a snowplow.

The poodle was lovingly nursed back to health and lived for many years, thanks to an unnamed German shepherd of the New York streets.

Researchers noted that a monkey, who was prone to seizures, was groomed by her group, including nonrelatives, twice as much as were her peers.

During "experiments" with food deprivation on capuchin monkeys, a monkey named Lucy was deprived of food while others in cages near her were fed. When she didn't lose weight, the surprised researchers were forced to conclude that the others had passed food through their cages to her.

Pie was a very independent female cat belonging to a family who lived in Wilton, Connecticut. Pie was the sort of cat that other people might say wasn't very loving or interested in her humans because she wasn't a lap cat and wouldn't cuddle.

However, Pie would prove them all wrong. When fire broke out in the house, she was the first to know. Her sharp nose told her of the danger and her instincts told her to flee. But Pie did not just escape as she could have. Instead she did something she would normally never do. She ran into the bedroom and jumped on the bed where the husband slept and pawed at his face until she woke him up. The family escaped and they all knew they had been saved by Pie.

"There are two refuges from the miseries of life: music and cats."

—*Albert Schweitzer*

"Cats have the most beautiful moves of any animal. They never think about it—they simply do it. I watch my cat all the time. I love to study him. If I could run like a cat, I'd be the greatest runner in the world."

—*Notre Dame halfback Eric Penick*

Sometimes an animal's devotion can affect our emotions even when human suffering cannot. Even Napoleon Bonaparte, the conqueror of Europe, was deeply moved by animal kindness. In his diary he wrote:

"Suddenly I saw a dog emerging from under the greatcoat of a corpse. He rushed toward us then returned to his retreat, uttering mournful cries. He licked the face of his former master and darted toward us again; it seemed as if he was seeking aid and vengeance at the same time.

"Whether it was my state of mind, or the silence of the guns, the weather, the dog's act itself or I know not what—never has anything on a field of battle made such an impression on me. I stopped involuntarily to contemplate the spectacle.

"That man, I said to myself, perhaps had friends, perhaps he had them in our camp, in his company; and yet he lies here abandoned by all except his dog. What is man? And what the mystery of his emotions? I had ordered battles without hesitation, battles that were to decide the fate of the army. I had seen, dry-eyed, the execution of movements that resulted in the loss of a great number of our soldiers. But here and now, that dog moved me to tears."

Napoleon was attempting to escape from his exile in Elba by ship when he fell overboard on a stormy night. A Newfoundland dog who belonged to one of the sailors rescued him.

Charlie, a pet raccoon, lived with the Margie and John Mertens family in northern Michigan during the 1960s. One night, while the people in his family of five all slept, Charlie's sensitive nose detected the acrid odor that told him the house was on fire.

Charlie rushed into the adults' second-floor bedroom and pulled furiously at the father's foot, trilling anxiously, until he'd awakened him. John then woke his wife and the two struggled down the hall through thick black smoke to their daughters' room as fire began to burst through the walls all around them.

As the parents reached the girls' bedroom, the floor behind them fell away, revealing that the downstairs was completely engulfed in flames—cutting them off from the second-floor nursery where the baby slept. The parents and their two girls reached the window as neighbors, who had seen the flames, were putting ladders up. The four barely escaped. But they were heartbroken, realizing the baby would be burned alive in the inferno.

The neighbors broke into the back of the house, just beneath the baby's room, battling flames and risking their own lives. They had barely gotten inside, protected by water-soaked blankets, when the stairway crashed to the burning floor, making rescue of the baby hopeless.

But just then a small masked form flew down from the floor above and landed at their feet. It was Charlie the raccoon—and in his mouth he was resolutely dragging the baby, who was burned but alive!

> "Any animal whatever, endowed with well-marked social instincts, the parental and filial affections being here included, would inevitably acquire a moral sense or conscience, as soon as its intellectual powers had become as well developed, or nearly as well developed, as in man."
>
> —Charles Darwin

When Anne Marie Schilling's children adopted an injured red-breasted robin, she never expected that it would remain with them as their devoted friend.

Many times she tried to return the bird to the wild, but it was devoted to the young boy and girl who had saved it. The woman relented to their pleadings to allow the bird to accompany the three on a long car-trailer trip, planning to get rid of the bird when they returned home. But the bird's heroism changed all that.

One night while Anne and her two children were asleep in the bunks of their trailer, the bird suddenly began shrieking. He landed by Anne's ear, shrilled, then flew furiously to the front of the trailer. When Anne got up, she realized what had alarmed the bird: A large idling tractor-trailer rig had pulled up next to their trailer so that rig's vertical exhaust pipe was spewing its fumes into an open window—and it was asphyxiating them! The driver was nowhere to be found. She closed the window, aired out her trailer and, with the robin's help, aroused her children from their potentially deadly sleep.

The next day, the family drove along a country road with the bird singing along with the radio, something he loved to do. Suddenly he became perturbed and flew madly around the inside of the car. Anne stopped the car. The robin flew out the car window, around the trailer, and back. The woman started to scold the bird—who chirped sadly at this undeserved punishment—but just then her children called, "Mother, come quick." She found that the trailer hitch had broken; had they continued driving down the road it would have meant disaster.

While waiting for repairs, the local mechanic was amazed by the robin's playfulness, devotion to his humans, and heroic deeds. The mechanic offered to buy the bird.

"Not for all the money in the world," Anne replied.

A Robin red breast in a cage
Puts all of heaven in a rage.
　　—William Blake

In 1991, as Michael Miller was snorkeling in the beautiful waters off the Big Island of Hawaii, he was suddenly caught in a riptide and carried a mile from shore. Though he swam as hard as he could, he soon realized that he was not going to make it back to land.

Michael closed his eyes and began to meditate. Despite his desperate situation, love and tranquillity filled his being. When he opened his eyes, he saw that he was surrounded by a group of sea turtles. The largest turtle swam directly under him, pausing so that he could easily cup his hand under the edge of its shell. The greenish brown shell felt soft and leathery to his touch.

The turtle swam Michael to shore and as they swam, Michael again felt flooded by tranquillity as he sensed the interconnection of all life. When they were close to the beach, he let go of the turtle which turned and headed back to sea.

Later, while driving home with his wife, Michael heard a report on the radio that there had just been a shark attack. A man had been killed in the same area from which the turtle had rescued him.

One day a pigeon flew into a boy's backyard. Though the bird wore the distinctive band of a homing pigeon around its leg, it showed no inclination to leave.

The boy began to feed the bird and spoke softly to it and the bird remained. It came to love the boy and their relationship deepened as they spent a year together.

One night the boy fell ill. His parents rushed him to a hospital 120 miles away. The next night, while recovering from his emergency operation, the boy heard a tapping on the window and he saw a pigeon standing on the windowsill, its feet almost frozen in the snow. The bird remained all through the night. The next morning the boy asked a nurse to open the window. The pigeon flew in. The identifying band on his leg proved him to be the boy's own faithful pigeon.

"[Concern for animals] is a matter of taking the side of the weak against the strong, something the best people have always done."

—Harriet Beecher Stowe

Ways to Return the Kindness

❧ *Did you know those plastic six-pack soda and beer can holders do not degrade for 450 years? And they can get stuck around an animal's neck or body sometimes killing them. Ask your grocer not to stock them. If you do buy them, cut the rings and the center diamond before throwing them away so they cannot choke any animals.*

❧ *Consider getting your next companion animal from the SPCA. Breeders, often called "puppy mills," too often keep their animals in crowded and unsanitary conditions. The so-called purebreds are much more likely to have health problems than so-called mutts. The adult cats used for breeding are bred constantly, which wears out their bodies, and mother cats are killed when their production goes down. Pet stores often get their animals from breeders.*

❧ *Ants can be humanely discouraged from entering a house by putting a line of cream of tartar on the floor. They will not crawl over it.*

☙ *Make sure garbage can lids are on tight. Otherwise, animals could raid your trash and get trapped in containers or cut on can lids. Many animals have gotten their heads trapped in those pyramid-shaped liquid yogurt containers trying to get the juice, and died as a result.*

☙ *When you travel, it is usually best to leave your animal at home. Have someone you trust walk your indoor dog four times a day. You wouldn't like it if you could only go to the bathroom twice a day either. Never leave an animal at the vet's. They could pick up diseases and they feel and hear the stress of the other animals, which is very hard on them.*

☙ *If you are traveling with your companion animal, most Holiday Inns accept pets in their rooms. Call 1-800-HOLIDAY to see if pets are allowed where you are going.*

A German shepherd named King was deeply attached to a man who died. Upon his death, the dog howled in mourning. On the day of the funeral, King bolted from the house and was not seen for three weeks. Finally, recalling King's attachment to their grandfather, the family went to the cemetery and learned from the groundskeeper that it had become King's habit to visit the grandfather's grave every day at two o'clock. There King moaned and howled. If anyone tried to approach him during his rite of private grief, King would growl.

Shortly before Abraham Lincoln's death, his dog wailed and howled and ran about the White House in a frenzy.

An eighty-two-year-old woman named Rachel Flynn was taking a solitary walk along the rugged coast of Cape Cod when she fell from a thirty-foot cliff to the isolated beach below. There she lay, in shock, wedged out of sight between large boulders.

As she lay there waiting for death, a seagull hovered overhead. It looked very much like "Nancy," the seagull she and her sister often fed at their home. Desperately she called out, "For God's sake, Nancy, get help!" The seagull flew off.

One mile away, the woman's sister was working in her kitchen when she heard a seagull tapping insistently at the window pane, flapping its wings wildly and, as she put it, "making more noise than a wild turkey." After shooing away the annoying bird for fifteen minutes, she began to wonder if the seagull was trying to communicate something to her.

She followed the gull outside. Nancy flew on ahead, stopping now and then to make sure the sister was following. Finally Nancy landed on the cliff edge and waited. The woman looked over the edge, saw her sister's predicament, and called for rescue assistance. The victim was taken to the hospital, bruised and with an injured knee. Both sisters are convinced that Nancy, the wild seagull, saved the woman's life.

Maria, a Husky dog, learned that by running around a cage full of parakeets and mice, barking furiously, she could terrify them all into a frenzy. This seemed like lots of fun to her.

Bingo, a pug dog who was very much enamored with Maria and almost never challenged her, entered the room. Upon seeing Maria's wild taunting of the caged animals, Bingo threw his body against the much bigger dog.

Only momentarily dissuaded, Maria again barked furiously, driving the mice and parakeets into pandemonium. This time Bingo barked loudly and lunged at her, knocking her aside. At this, Maria ran off, leaving the caged animals in peace.

This story is retold in the book *When Elephants Weep*, where authors Jeffrey Moussaieff Masson and Susan McCarthy comment: "It seems undeniable that he wanted to stop her aggression and make her behave better toward the other animals."

> "Any religion not based on a respect for life is not a true religion. Until he extends his circle of compassion to all living things, man will not himself find peace."
>
> —Albert Schweitzer

K oko is probably the most famous of the signing gorillas and there are many stories of her intelligence, humor, and kindness. Her love for kittens is particularly touching. When Koko was first introduced to some visiting kittens, she signed, "Visit love tiger cat."

At her request, the childless gorilla was allowed to have her own kitten to care for. Koko's kitten, like all kittens, tended to put her claws out sometimes, but instead of being angry, Koko just signed, "Cat do scratch Koko love."

One day, Koko's guardian and teacher, Penny Patterson, brought Koko the terrible news that her beloved kitten, "All Ball," had been killed by a car. At first, Penny misinterpreted Koko's silence as not understanding. Penny left Koko alone for awhile. Ten minutes later she heard Koko making the gorilla distress call—high pitched, staccato hoots. Penny knew then that Koko had understood. A few days later, in a conversation about the departed "All Ball," Koko was asked if she wanted to talk about her kitten. Koko sadly signed "Cry blind sleep cat."

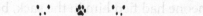

The first words Koko taught her new gorilla friend Michael to sign when he arrived at her compound seemed to indicate just what she wanted from their relationship. The first word was "Koko." The second word was "tickle."

Two neighbor cats were close companions and were often seen together, scampering along fence tops or blinking sleepily in the sun.

Then one cat disappeared. The remaining cat meowed and meowed, insisting on coming along in the car as his friend's family prepared to search for the missing cat. They tried to throw the distressed cat out of the car several times but he kept hopping right back in. Finally they relented and allowed him to accompany them. He meowed, and with his body and intent gaze, seemed to indicate which direction they should go.

Since they had no clear idea of where to look anyway, the humans decided to drive in the directions the cat indicated. They wound up driving slowly along the waterfront, but there were no cats in sight.

Suddenly the tense cat leapt from the car and rushed over to a burlap bag which he furiously tried to open. When the people got out of the car and opened the bag, there was their lost cat, injured but still alive.

Someone had tied him in the sack, beaten him, and left him to die at the waterfront. But his friend had sensed his silent call for help and led his human family to the rescue. The injured cat recovered and the two remained inseparable friends.

His battalion was remorselessly pounded by German shells, gunshots were exploding all around him, but "Cher Ami," a military homing pigeon during the First World War, bravely flew off into the skies with an important message wrapped around his leg.

The American soldiers watched as their little bird took off but then despaired as they saw the Germans fire upon him. The pigeon fell to the ground. Then someone yelled out, "Cher Ami, go home!" Incredibly, the fearless little bird took off again.

Again, a rifle slug painfully pierced his breastbone as he strove to fly. Yet another bullet tore off his right leg and a fourth took out one of his eyes. Showing immense courage, Cher Ami still tried his best to fly onward. Amazingly, the little bird struggled on and made it to the roof of his destination—Allied headquarters. Exhausted and dangling a leg, he collapsed on the rooftop as the soldiers removed the band from the torn leg, which was attached only by a ligament. The desperate message was that the American forces were being bombed, in error, by their own side. The firing immediately ceased and his battalion was saved.

Cher Ami recovered and lived until just after the end of the war. After his death, his body was stuffed and placed at the Smithsonian Institute where it still can be seen. Beneath the mounted body of the little one-legged soldier is the inscription: "The feathered hero of World War I."

"I could not have slept tonight if I had left that helpless little creature perish on the ground."

—Abraham Lincoln to friends who upbraided him for delaying when he interrupted a journey to return a fledgling to its nest

A fallen sparrow was lying helplessly in the middle of a crowded Italian street. It was soon surrounded by other sparrows trying to carry it to safety, despite the heavy traffic in the area.

A man got out of his car and waved to the other drivers, thereby stopping the traffic. Soon traffic was at a standstill. Slowly, with great effort, the little birds managed to lift their wounded companion and carry him to the side of the road. They rested on some grass for a moment. At last, with great coordinated effort, they propped up the injured sparrow and flew it over a wall and into a garden.

The sparrow's rescue was in sharp contrast to a study of human motorists' behavior in Europe where hundreds of people sped by a faked accident, none stopping to assist. Observing the behavior of the birds, which are considered a delicacy in the region of Italy where the incident took place, one man remarked: "In these little feathered creatures there is something more than a couple of ounces of meat with which to season a plate of cornmeal."

A hearing dog named Cookie awakened her two human companions one night. Both were deaf from birth and had recently had a baby.

They immediately rushed to the bed of their infant daughter who had been suffering from a chest cold. She was choking and turning blue. Her parents immediately treated her and she soon made a complete recovery.

"The dog woke us!" the father said later. "She jumped on the bed and woke us up! Cookie knows the difference between regular cries and cries of panic or pain."

Hearing ear dogs learn to respond to doorbells, ringing telephones, alarm clocks, oven timers, tea kettles, and knocking at the door. They can be trained not to jump on their human's bed, but they are also trained to break that taboo when the smoke alarm goes off and the person is sleeping. They are so indispensable to deaf folk that the IRS has ruled that their expenses are deductible.

During a scientific study in March 1973, while projecting some film for a documentary, a Russian entomologist named Dr. Marekovsky discovered something unusual. He had been collecting footage on insects for a long time and finally noticed an odd behavior he had not seen before: a scene in which three Amazonian ants were extracting a splinter from the side of another ant.

The doctor ant worked carefully as other ants in the colony formed a circle around the patient and doctor ant to protect them while he worked. And, as far as Marekovsky could tell, the patient was never even billed for the surgery.

Chimpanzees have been observed to bring food to their old mothers who can no longer climb for it.

D o you believe a cat can psychically sense a human's distress? The English magazine *Tomorrow* offers this story:

Bill the cat stayed home while his human was away on a trip. But the man was injured in a railway accident and died a few days later in a hospital. At the burial, the man's brother was shocked to see Bill in attendance. The faithful cat had journeyed to the distant hospital grave site, looked down upon the coffin, and then sadly returned home.

> "We are part of the earth and it is part of us. The perfumed flowers are our sisters; the deer, the horse, the great eagle, these are our brothers. The rocky crests, the juices of the meadows, the body heat of the pony, and man—all belong to the same family."
>
> —Chief Seattle

Chips was the first dog to be sent overseas for active duty during World War II. He served in both Europe and Africa, where dogs were used for reconnaissance patrols, security, and village-clearing missions. Their keen noses—forty times more sensitive than a human's—enable them to find buried mines, trip wires, ammunition, and enemy soldiers. One night in Germany, Chips assaulted a concrete gun-emplacement by himself, captured one man, and terrified four others into surrendering. He was awarded the Purple Heart and the Silver Star, but the army later rescinded these awards because he was a dog. His own company amended this by creating a medal just for him, and gave Chips a battle star and campaign ribbons. He retired to civilian life after the war.

"Here lies the body of my good horse, The General. For twenty years he bore me...and in all that time he never made a blunder. Would that his master could say the same."
—President John Tyler

A young rhinoceros was trapped in mud at an African animal park. The day was hot and sticky and the mud had dried just enough that the little rhino couldn't raise its legs to get free. It cried for help and its mother came, first checking to see if the youngster was hurt in any way. Seeing that her baby wasn't injured, the mother didn't seem to understand the problem and walked away. But still the little rhino bellowed for help.

Then an elephant approached. Although certainly not a relative, and not even of the same species, the elephant tried to pry the baby rhino loose by getting its tusks under it and lifting it out. As he struggled, the rhino's mother came back. Thinking the elephant was trying to hurt her little one, the mother rhino charged the elephant, threatening to gouge it with her horn. The elephant withdrew.

A bit later, again temporarily abandoned by its mother, the baby called out, so the good-hearted elephant returned and tried again. For several hours, despite the danger of being repeatedly charged by the violent mother, whenever the mother left, the elephant came back and tried to help the rhino baby. At last the elephant gave up all hope of rescue under such hostile circumstances and left.

The next day, as park authorities prepared to attempt a rescue, the mud dried, and the baby struggled free by itself.

When lions are shot with anesthetizing darts, other lions have been seen to pull the darts out of their unconscious friends' bodies with their teeth.

Even bats have been known to help each other out in times of need. Gerald Wilkinson, a scientist who studies bats, has found that in lean times a bat who has not been able to find food may be fed by a friend via regurgitation. Because the bats he was studying cannot live more than two days without food, this help may mean life or death to the recipient bat.

The kindness of animals goes far beyond deeds of heroism. Their natural good-heartedness often takes much subtler forms. For example, scientific studies have shown that stroking and talking to an animal can be good for you. "We've found that blood pressure and heart rate go down considerably while a person is petting a dog. It has a deeply calming effect, like prayer or meditation," said Kenneth White, education coordinator of San Francisco's SPCA.

Many medical experts agree that animal companionship can improve your health and even extend your life expectancy. At a meeting of the American Heart Association, for example, it was reported that reductions in blood pressure have even been noted in subjects who are doing no more than simply talking to birds or watching tropical fish.

University of Pennsylvania biologist Ericka Friedman stated in her research on heart attack patients that one statistically significant factor in recovery was sharing one's life with an animal companion. This was significant no matter what the degree of illness. Even studies excluding dogs, who exercise their human companions on their daily walks, showed that patients with pets had a much higher rate of survival.

One day, a young impala—a graceful African antelope—was drinking by a river, when he was grabbed by a crocodile lurking beneath the muddy surface. The crocodile started to pull its victim underwater.

A nearby hippopotamus witnessed the attack and chose to defend the victim. The furious hippo—capable of biting a crocodile in two—charged; the crocodile released the impala and swam away.

Then the hippo gently nudged the impala to the shore with its nose and onto higher ground. Not content to simply rescue it, the hippo stood guard against other predators for fifteen minutes. Then the two-ton animal began sniffing it and gently licking its wounds. Twice the hippo took the impala's head into its mouth in what seemed to be an attempt to get it to stand. But it was mortally wounded. Since hippopotamuses are strictly vegetarian, its interest in the dying impala was purely altruistic.

This story is all the more amazing because it is completely documented. A *Life* magazine photographer, Dick Reucassel, was present for the incident and recorded it all on film.

As I did the research to substantiate this amazing story, which had been told to me by a friend who remembered it from childhood, I had to go through many old copies of *Life* and other resources to find the original article. To my astonishment, in the course of looking I found yet another hippopotamus-saves-impala story. This one was reported by Harry Erwee, a researcher at the Masuma Dam in Hwange National Park, Zimbabwe, and substantiated by other witnesses.

Nine wild African dogs were pursuing an impala. Half the pack, the witnesses could see, was waiting in the bushes on the other side of some water while the first half chased the now tiring impala toward their waiting companions. A hippopotamus swam toward the weakening buck and prevented it from going to the other side where the dogs were laying in wait; instead, it nudged it in the opposite direction. As the exhausted impala lost energy, the hippo would nudge it forward. When the duo finally reached safe ground, the impala was too exhausted to make it onto dry land, so the hippo nudged it with its snout. The impala began shivering violently. Too weak to make it on its own, it was on the verge of collapse. The hippopotamus then opened its gigantic mouth (so large it could encompass the small antelope without actually touching it) and repeatedly breathed life into the animal until it regained its strength and ran off.

In an article he wrote for BBC *Wildlife* magazine, African researcher Harry Erwee described the incident as "awe-inspiring." I have nothing to add to that.

"How are we to build a new humanity? Reverence for life. Existence depends more on reverence for life than the law or the prophets."

—Author unknown

According to Rae Phillips of New Mexico, Buddy, his cat, "is a real environmentalist. Every day he picks up plastic bags, candy wrappers and any other litter behind our house and brings it to us to dispose of. Sometimes it's difficult for him to get the litter over the fence since he often brings home items like garbage bags with other trash inside!"

Rae supplied pictures to *Country Magazine* of his amazing trash-carrying cat. Rae says his cat will even clean up trash at campsites when the family takes a trip in their motor home.

※

A lonely orange cat was about to be destroyed at a Chicago animal shelter. He had just twenty minutes of life left when—just like in old Hollywood movies—he was discovered by show biz people. Renamed "Morris," he went on to become the world's most famous cat, thanks to cat food commercials. Up until his death many years later, he lived in luxury and had armed guards to protect him from catnappers.

J ane Goodall tells the story of David Greybeard, one of the chimpanzees she studied in Gombe, Africa. One day, after lying peacefully close by David Greybeard as he slept, Jane picked a ripe palm nut off the ground and held it out to him as a gift. The chimpanzee let go of the nut, but, for an instant, closed his fingers around her hand.

"He rejected the gift but not the giving," said Goodall. "His message had no need of words—it was based on a far older form of communication and it bridged the centuries of evolution that divided us."

"We must never permit the voice of humanity within us to be silenced. It is humanity's sympathy with all creatures that first makes us truly human."

—Albert Schweitzer

Gombe chimpanzee researcher Geza Teleki was following a group of tree-climbing chimps one morning when he realized he had forgotten his lunch. Hungry, Teleki tried to knock fruit out of a tree with a stick but, not being much of a climber and the fruit being very high up, he couldn't get any. Fruit gathering proved much harder than the chimps made it seem.

Discouraged, he realized that he was probably just going to go without lunch that day. After a short while, a chimp named Sniff who was gathering some fruit from nearby trees, approached the man and, to Teleki's surprise, offered some to him.

"Knowing all truth is less than doing a little bit of good."

—*Albert Schweitzer*

It was an elderly woman's worst nightmare come true. The man she had let into her home because he said he was a meter reader violently knocked her to the ground. She had no weapons and her strength was no match for his. She knew he could kill her.

Suddenly an animal leapt onto the burglar's shoulders and sank its claws into the back of his neck. Terrified, he let go of the woman and ran, in shock and pain, from her home. The woman locked her door and was safe.

She had been saved by her cat.

> "This is my first friend. I love him and he loves me. I didn't know what love was until I was given this bird."
> —A multiple murderer speaking about his parakeet provided by a special program for resocializing prisoners through the use of animals

According to Science News magazine, *a companion animal is, for many lonely juvenile delinquents and emotionally disturbed youngsters, the sole love object in their lives.*

Human (if not humane) researchers decided to assess "maternal instinct." So they tested mother rats by placing their babies on the other side of an electrified grid. Every time a mother rat tried to cross the grid to reach her babies, she would receive electric shocks. But the pain of this crossing did not discourage the mother rats, who were as devoted to their offspring as any other mother. Wanting to nurture their babies and protect them from the painful situation created by the researchers made the mother rats resolute.

The courageous little mothers repeatedly crossed the grid, even though they received many shocks, to retrieve their babies. Some mothers even rescued babies that were not their own. One daring mother rat carried fifty-eight babies across the electric grid, receiving shocks all the way, before researchers ran out of infants with which to test her.

> *"The greatness of a nation and its moral progress can be judged by the way its animals are treated...I hold that the more helpless a creature, the more entitled it is to protection by man from the cruelty of man."*
>
> —Mahatma Gandhi

Charles Darwin wrote an account of an incident at a zoological gardens in which a zookeeper was attacked by a fierce baboon. The baboon charged the man and knocked him off his feet. The baboon weighed almost as much as the man but his powerful arms made him many times stronger. The man lay helplessly upon the ground. Stunned, and momentarily unable to defend himself, he expected the baboon to finish him off. Then a small monkey acted.

This tiny monkey shared the cage with the baboon and was terrified of him. But the monkey had long regarded the man as his friend. Putting aside his own safety, the little monkey charged and bit the baboon. Bravely he battled the baboon who was many times his size. Due to the monkey's furious attacks, the baboon was distracted for an instant, just enough of a chance for the zookeeper to make it to safety. He crawled out the door and slammed the cage door behind him before the baboon could follow. His wounds were deep and he bore lifetime scars on his neck but he was alive. His little hero monkey friend also survived.

Ways to Return the Kindness

❧ *For natural flea control, try sprinkling a little garlic powder and brewer's yeast on your pet's food. It's good for you, too.*

❧ *Never leave a dog in a parked car on a warm day. The temperatures can reach 10° to 20° warmer than the outside. Also, your dog could be stolen by a "buncher" who sells stolen pets to labs for horrendous experiments on live animals. It is conservatively estimated that two million dogs are stolen each year. 300,000 dogs and 100,000 cats are used in laboratory experiments each year.*

❧ *Clean up litter where you find it to protect wildlife.*

❧ *Whole bay leaves placed in corners around the house will discourage cockroaches.*

❧ *Don't use hummingbird feeders. Unless they are cleaned every other day, the sweet liquid isn't fresh and causes intestinal problems in hummingbirds, which is killing them off.*

❖ *Plant trees and berry bushes to help offset the destruction we are causing the environment and to create habitats and food for birds.*

❖ *At weddings, throw birdseed instead of rice. When the wedding is over, birds arrive and peck at the rice. But uncooked rice swells in avian stomachs, killing them.*

❖ *It's a nice idea to keep dry pet food and water in your car in case you see a starving animal that you can't catch. Leave the food where you last saw the hungry animal; it will find the food when you leave.*

Ginny, "The Dog Who Rescues Cats," is the subject of a book by the same name. She once pulled her human guardian, Philip Gonzalez, into a building that was under construction, ran up to the second floor, and banged on a pipe with her paw.

Philip tried to restrain her, but the little dog would not be dissuaded. She ignored Philip's human concerns about dismantling private property. The determined little dog tried to shake the pipe loose. As Philip attempted to get her to stop, she pawed at the pipe again and again until she knocked it down. There in the pipe Philip found a litter of tiny starving kittens no more than a week old. Someone had thrown the kittens away and left them to die. They were tiny and miserable and covered with fleas but alive. Somehow, the dog had heard their tiny desperate mews and found them.

Philip and Ginny took the kittens home. They were so young that they had not yet opened their eyes. Feeding them kitten formula and gently removing the fleas, Philip nursed them all back to health. Every single kitten survived.

Ginny has become famous for leading Philip on nightly hunts for cats who are hurt and hungry. She finds animals who need help and Philip brings them home to feed and heal. He usually places them in good homes but couldn't resist keeping two kittens for himself.

"I have been scientifically studying the traits and dispositions of the 'lower animals' (so-called), and contrasting them with the traits and dispositions of man. I find the result profoundly humiliating to me. For it obliges me to renounce my allegiance to the Darwinian theory of the Ascent of Man from the Lower Animals; since it now seems plain to me that that theory ought to be vacated in favor of a new and truer one, this new and truer one to be named the Descent of Man from the Higher Animals."

—Mark Twain

Gerald Coward, a Los Angeles photographer, was walking through a canyon near his home one day, and met a coyote. Perhaps because it reminded him so much of a dog, Coward decided to walk with it.

The coyote showed up the next day and they walked together again. This continued for two and a half years. The coyote would be there at the appointed time and the two would walk and play together. One day there was a huge fire in the canyon, and, tragically, Coward never saw his friend again.

"For whatever happens to the beasts soon happens to man. All things are connected. This we know. All things are connected like the blood which unites one family. All things are connected. Whatever befalls the earth, befalls the sons of the earth. Man did not weave the web of life, he is merely a strand in it. Whatever he does to the web, he does to himself."

—Chief Seattle

A sparrow crashed into a chimpanzee cage at the Basel Zoo. One of the chimps scooped the bird up in its hand.

A zookeeper who was watching this expected to see the animal devour the bird. But instead of eating it, the chimp just held the bird tenderly and contemplated it.

Other chimpanzees became curious and came over to see the bird. It was delicately passed from chimp to chimp. Each chimpanzee examined the little creature, holding it gently in its great ape hands, taking obvious care not to hurt it. Finally, realizing that the bird could be better cared for by humans, one chimpanzee brought the bird to the front of the cage and carefully handed it to the startled zookeeper.

"You miss who we are, thinking about us the way that you do."

> —*The reply of an ex-convict who works with elephants, when asked what he thought was the one statement these animals would communicate to man if they could.*

The female trapeze artist swung through the air, did a somersault, and reached out to be caught by her partner. Although their fingers brushed each other, their hands did not quite connect and she fell.

There was no net. The audience fell silent in horrified anticipation. All eyes were upon her as she plummeted to the circus floor.

Suddenly, in what must be one of the most incredible moments in circus history, a trained horse broke formation and ran to the spot where the trapezist would strike the hard earth. Instead, the woman fell upon the horse, knocking it down as they both tumbled to the ground. The woman was unconscious but uninjured and the horse got up and trotted away.

> *"The monstrous sophism that beasts are pure unfeeling machines, and do not reason, scarcely requires a confutation."*
>
> —*Percy Bysshe Shelley*

High school boys often get into playful and, sometimes, not-so-playful fights. In one school, a large thick-furred collie started placing itself between the fighting boys so they could not strike each other.

The dog never bit or growled and its thick coat ensured that it wouldn't be hurt by any unexpected blows. Soon the collie started patrolling the school grounds just before school started and stayed until the grounds were empty. The boys laughed so hard when the collie pushed its way between them that they often forgot why they were fighting. And of course, they would often start "fights" just to have their furry companion come save them.

"In the relations of man with animals…there is a whole great ethic scarcely seen as yet, but which will eventually break through into the light and be the corollary and complement to human ethics."

—*Victor Hugo*

In *Kinship with All Life,* J. Allen Boone, the great-grandson of Daniel Boone, describes how his theories about compassion and communication with animals were put to the test.

On a camping trip, he returned to his cabin to find ants everywhere devouring his food. He went to get ant poison and a broom when he realized that, rather than simply killing the ants, he should at least try the animal communication techniques he had been recently studying.

Boone had learned to communicate with dogs and horses and even snakes. So why not at least try it on ants?

Studying the ants to determine what he could like and respect about them (the beginning of the psychic bond which leads to communication), Boone settled on diligence and cooperation as the likeliest traits. After praising them mentally, he told them that he could kill them but instead was humbly requesting that they leave his food alone because he sorely needed it. He left for awhile and upon his return, all the ants had graciously departed.

I had occasion to try Boone's technique myself once when I found cockroaches in my home. I confess I tried boric acid and roach motels first, with limited success, before I remembered reading of Boone's unorthodox methods.

I sat down to accomplish the first step of finding something admirable about the cockroaches. After about fifteen minutes of uneasily watching them, I confessed to them that I could see nothing I liked but that if they would leave, that would be very admirable and I promised to say good things about cockroaches whenever asked.

I even wrote them a letter, the next step recommended by Boone, and tucked it into the garbage, which seemed a likely mailbox—and where it would be away from the prying eyes of any friends who might find me completely crazy. The next day, to my astonishment (I hadn't much faith in Boone's theories either) the cockroaches were gone and never returned, although I lived there for two more years.

> *"The original instructions of the Creator are universal and valid for all time. The essence of these instructions is compassion for all life and love for all creation. We must realize that we do not live in a world of dead matter, but in a universe of living spirit. Let us open our eyes to the sacredness of Mother Earth, or our eyes will be opened for us."*
>
> *—David Monongye, Native American elder*

One day, a man set out fishing for flounder off the New England coast with his Irish setter, Redsy, who loved these fishing trips. But this day Redsy stalwartly refused to get in the boat, barking stubbornly in response to her human's stern entreaties. The weather was perfectly calm, but the man had never seen Redsy so adamant before. Watching regretfully as fifty other fishing boats set out for the flounder banks, he canceled the day's fishing trip because he had great respect for Redsy's intuition.

Within an hour, a powerful storm thrashed the New England coast. Up to forty foot waves crashed upon boats and smashed the coastal cottages. More than six hundred people died in the great hurricane of 1938, but Redsy's warning had saved her fisherman friend.

> *"The fate of animals is of greater importance to me than the fear of appearing ridiculous; it is indissolubly connected with the fate of man."*
>
> —Emile Zola

One hot sunny day, a woman who had rescued a starving, mistreated calf took it back to her horse corral. There, after getting it rehydrated as best she could, she settled down to pray for its life.

As the limp little animal lay in her arms, she noticed that the horses in the corral were behaving strangely: they stood clumped together nearby and seemed to be watching the sick animal. This struck her as odd but she did not concern herself because she was concentrating on the calf.

Then she noticed that as the sun moved, the horses moved too. Later the position of the sun changed and the horses moved again. They were creating shade and protecting the helpless little animal from the heat. They, too, had been giving care to the stricken animal in the only way they could.

The woman stayed up all night with the calf, massaging it and telling it that it was loved. The calf was very weak and struggled for its life during the night. Finally its strength came back and it survived. It was later named "Susie" after Susan Hayward in the movie, *I Want to Live*.

In Euless, Texas, a woman was doing her housework one day when she saw a scruffy puppy scratching at her back door. She shooed it away because she wanted to get her children a purebred puppy for Christmas. Later she went outside and was confronted by the pitiful sight of the starving puppy attempting to eat a mop. Her heart broke and she fed it. Her children soon saw the puppy and fell in love with it, refusing to accept any other dog. She didn't know it then but it was lucky for her she kept that dog!

One day, her little boy, two-year-old Randy, wandered away from home. His desperate parents searched everywhere, but they could not locate him. The police were called in and after two hours they still could not find the child.

Three-quarters of a mile away, motorists were stopped by what they thought was a mad dog in the middle of the street. On closer inspection, it was discovered that the "mad dog" was that very puppy, now big and grown up. The dog, now named Ringo, was charging at cars and refusing to let them pass because little Randy was playing in the middle of the street.

When the toddler had wandered away from home, the faithful dog had trotted protectively behind him. Every time a swiftly moving car was about to run down little Randy, the dog would charge the cars, even leaping on their fenders. Dog and boy were finally coaxed to the side of the road, and Randy was returned safely to his grateful parents.

•᎓• ❀ •᎓•

Ann Cain, professor of psychiatric nursing at the University of Maryland, researched family dynamics. Studying sixty households she discovered that families' attitudes changed once an animal was introduced. Many families experienced "more closeness, more time playing together, and less arguing after they had gotten their pets." One resourceful mother used the family dog to put an end to arguments. "Stop fighting," she'd say. "You're upsetting the dog."

At the Ohio State Hospital for the Criminally Insane, a prisoner found an injured wild bird and tried to care for it. In caring for the bird, he began to heal some of the suffering within himself.

The prison psychologist, David Lee, noticed the man's improvement and also realized that many of the other men on the ward were interested in the bird's welfare too. He got permission from the supervisors to buy a fish tank and two parakeets for the men.

All went well until it was discovered that one of the parakeets was missing. A frantic search of the ward ensued and finally it was determined that one of the hardened "criminally insane" inmates, who was being transferred to another facility, had taken the bird. The parakeet had awakened a gentleness in his heart that years of punishment had not yet destroyed. Unwilling to leave the parakeet behind, this violent criminal had hidden it in his duffel bag hoping to sneak it into the new prison.

Thereafter, the prison supervisors realized they had a new way to reach their prisoners. They decided that the men would have the right to earn pets of their own, and many more pets were introduced to the ward.

Ken Wilson was busy trying to train a horse and didn't pay any attention when his dog Taffy started barking at him. Ken thought his three-year-old son Stevie was safe where he had left him inside the house. But Stevie had wandered off and stumbled into a pond, where he had sunk to the bottom.

Taffy barked madly at Ken and finally, in desperation, started nipping at the horse's heels. At last Ken realized something was really wrong and followed Taffy. The dog led him to the pond. Ken's heart raced when he saw his son's jacket floating on the surface.

Rescued, but unconscious, little Stevie lay in a coma as the family prayed for six hours. The boy finally regained consciousness and the first thing he saw was his happy hero dog sitting at his bedside, keeping watch.

"When you love every creature, you will understand the mystery of God in created things."
—*Fyodor Dostoyevsky*

The world has heard many amazing stories of Seeing Eye dogs, but how about a Seeing Eye cow? This incredible guardianship was first reported in *Fate* magazine.

As the Reverend O. F. Robertson of Tennessee had begun to lose his eyesight, his cow, Mary, began nudging him around the hilly farmland with her nose, making sure he didn't bump into things. As his eyesight worsened, Mary's diligence increased until she would accompany the reverend wherever he went. Eventually almost totally blind, the reverend depended on Mary the cow as his only help and until his death, the two could often be seen making the rounds together, Mary steering him across the countryside.

A man who works with a single elephant in India is called the elephant's mahout. If the mahout trains his elephant with kindness, a powerful bond of love may form between them. If the mahout dies, it is not uncommon for the elephant to become depressed and refuse to eat until it also dies. If the elephant lives, it usually takes two or three years to get used to a new mahout. One Cambodian elephant recognized his mahout's grave a half-century after his death.

The Earl of Southampton, Henry Wriothesley, who lived from 1573 until 1624, was imprisoned in the Tower of London during the reign of Elizabeth I. Utterly alone with nothing but stone walls around him, he contemplated the miserableness of his fate. His fear was deep because many people did not leave the dreaded Tower of London alive. No friends or relatives could help him there.

Suddenly he saw a familiar form. Was he going mad, too? But it was not a hallucination. It was his own sweet house cat come to cheer him and sit by his side in his hour of greatest need. He could scarcely believe his eyes. The faithful animal had climbed down a prison chimney to be with him. After the Earl was released by King James I, he had a portrait painted of himself with the little black-and-white cat in the background.

"I've loved animals all my life—their friendship has always meant as much to me as any of the friendships I've had in my life."
—*Candice Bergen*

It was August 1936, in Fort Benton, a prairie town in Montana. A casket was delivered to the train for transport to a distant burial site. A single mourner accompanied the body—a mongrel dog by the name of Shep. As the casket was loaded, the dog attempted to follow but was shooed away. He whined pathetically as the train pulled away taking the body of his only friend with it.

Inconsolable, the dog sat down on the railway platform to wait for his friend's return. In the end, Shep would wait for five and a half long years before he himself died by the very tracks where he had waited so patiently.

But the effect of Shep's devotion lived on. His story was carried in Ripley's *Believe it or Not*. So many people inquired about the dog that a conductor named Ed Shields wrote Shep's story in a booklet. When sales of the booklet topped $200, Shields decided to give the money to the Montana School for the Deaf and Blind. The counselors were able to buy all sorts of toys, games, and pretty clothes for the children that state money would not provide. Christmas became merry at the children's school, thanks to Shep.

People across the nation heard the story of Shep's devotion. Their hearts opened and they wanted to share. Over the years, the Shep Fund, as it came to be called, raised over $55,000 in donations.

Because of Shep, the children received the luxuries that made them feel that they were special and worthwhile. They were able to vacation on a ranch and have some real fun. Self-esteem rose in the children and instead of just one or two qualifying for college, up to 40% of the graduating classes began to qualify. Shep's love and his heartfelt devotion to his human friend inspired so many that his legacy endured and continued to uplift the lives of others for many years.

"There are, I know, people who do not love animals, but I think this is because they do not understand them—or because, indeed, they do not really see them. For me, animals have always been a special part of the wonder of nature—the smallest as well as the largest—with their amazing variety, their beautifully contrived shapes and fascinating habits. I am captivated by the spirit of them. I find in them a longing to communicate and a real capacity for love. If sometimes they do not trust but fear man, it is because he has treated them with arrogance and insensitivity."

—Pablo Casals

Christopher Mazzella was an eight-year-old autistic boy who spoke only in robotic mimicry of the voices he heard around him. The only food he would consent to eat was crackers and cheese.

His desperate mother took him to Patricia St. John, author of *The Secret Language of Dolphins*, to see if he could be helped. St. John arranged for Christopher to swim with a dolphin.

The first thing Christopher did was hoot and chortle. The dolphin hooted back. Christopher and the dolphin played and swam. Even just one day of swimming with a dolphin changed Christopher. For the first time he ate healthy food.

For the next two weeks he swam with a dolphin every day, and this helped Christopher come out of his isolated world. Three years later, Christopher and his mother have a lot to be grateful for. He eats many different foods and likes people more. "He will always be autistic," his mother says, "but he is reachable now."

∴ ❀ ∴

Dolphins and whales sometimes attempt to help their companions when men have injured them by biting harpoons or biting through nets in which their friends are trapped.

From barbaric ancient Rome comes this classic account of affection between man and beast. This famous incident was witnessed by the Roman citizenry and is described in the book *The Attic Nights of Aulus Gellius.*

In the Great Circus, wild beasts were exhibited to the people. There were many savage beasts, brutes remarkable for their huge size and unusual ferocity. Beyond all others, the size of the lions caused excitement, and one in particular surpassed all the rest because of the huge size of his body, his terrifying and deep roar, his muscles, and his mane flowing over his strong shoulders.

Many slaves were brought in, having been condemned to death. One was the slave of an ex-consul. When the great lion saw him from a distance, he stood still as if amazed, and then approached the slave slowly, in gradual recognition. Then, wagging his tail like a pet dog, the lion came close to the terrified man and gently licked his feet and hands. Then the man and lion greeted each other in joyful reunion.

The emperor Caligula stopped the performance and wanted to know why the lion had not eaten the man. The slave, Androcles, told how he had run away from his cruel master and into the lonely desert where he had hidden in a cave. He had decided to choose death, if he had to, rather than live the life of an abused slave. Then a lion entered the cave with a wounded paw, moaning in pain. "The lion," Androcles said, "approached me with gentleness, and lifting up his foot, was evidently showing it to me and holding it out to ask for help." In his paw was a huge splinter which Androcles removed and treated. "Relieved by my attention and treatment, the lion, putting his paw in my hand, lay down, and went to sleep."

Androcles and the lion lived in the cave together for three years. They became such good friends that the lion hunted for both of them. One day, Androcles was captured, sent back to Rome and condemned to death in the arena. After hearing this incredible story, Caligula, one of the cruelest emperors ever to live, decided to put their fate to a vote of the people. The people voted to free them both and Androcles and the lion remained friends for life.

Elsa the lion was raised by Joy Adamson and made famous in the book Born Free. *For years after being released into the African wilderness, Elsa would return to greet her human friends, even bringing along her cubs.*

At a very low point in her life, a woman saw no way out and decided to commit suicide. As she sat on her bed weeping and wondering what method to use, her cat jumped on her and began licking her tears away. Realizing that when another creature loves you there is always hope, the woman decided to live and ultimately went on to write this book.

Thank you, Yoko.

In Glasgow, Scotland some hospices let cats stay with terminally ill patients. The cats lie quietly with the patients, lessening their fear of dying alone.

Ways to Return the Kindness

☙ Keep your local SPCA phone number with you at all times so you can report an injured animal quickly. This can be critical in sparing an animal great suffering.

☙ Avoid circuses that use animal acts. Those adorable gestures that humans find so interesting are unnatural and painful to perform. People would not enjoy their antics so much if they saw that the tricks are often the results of training by beatings or electrocution. So-called dancing bears frequently walk upright only because their front paws have been painfully burned. The Cirque du Soleil and Pickle Family Circus use talented people instead of animals.

☙ When flying, call the airlines in advance and see if you can take your animal with you in an onboard carrier. Temperatures are not controlled in the cargo areas and pets can die painfully. If you must put them in cargo, fly at night in the summer or by day in the winter. Never fly with animals where there are temperature extremes and always choose a direct flight.

❧ *You can discourage ants by washing surfaces with equal parts vinegar and water.*

❧ *You can do both yourself, your family, and the animals a big favor by trying to eat more healthful vegetarian food. When her older brother, Michael, teased Janet Jackson about her weight, she followed a vegetarian diet and lost the baby fat. Going totally vegetarian (vegan) reduces your risk of heart attack by ninety percent. Many celebrities—even Madonna—are vegetarians! Vegetarians are, as a rule, thinner than non-vegetarians. A vegetarian diet makes a lot of sense from an environmental perspective. According to John Robbins in his book* Diet for a New America, *it takes fifty-five acres of decimated rainforest to make a quarter-pound hamburger. The amount of water used to make just one of these burger patties is equivalent to the amount used by a family of four for one month.*

L arry couldn't seem to shake a drinking habit that often kept him up all night going to bars. His life sank deeper and deeper into bouts of drunkenness and blackouts; his drinking even once led to his hospitalization. Larry was a good man, but so addicted that he needed a miracle to help him stop drinking.

One day, a scruffy stray dog found him. The man took the dog in and soon came to love the dog whom he named "Homer." Recognizing the great affection between them, the man now felt cruel when he went out to bars at night because he knew Homer was lonely in his apartment, just waiting for him. The man began to return home from work early so he could take his dog for long walks instead of going barhopping.

Larry joined Alcoholics Anonymous (AA). He started to enjoy life again and found that going home to someone who loved him made all the difference. Three years after joining AA, he received a three-year sobriety pin. Giving credit where credit was due, he attached the pin proudly to Homer's collar.

A man was relaxing with his Alsatian dog at the local tavern. Suddenly the dog became agitated, doing everything possible to attract attention; running in circles, howling, tugging at his human's clothes.

The dog tried desperately to drag his friend toward the door. The man threw his annoying dog outside and tried to enjoy his drink. But the dog climbed back in through a rear entrance and resumed tugging at the man's clothes. Tired of fighting, the man followed his dog outside.

Two minutes later, the entire tavern collapsed because of structural damage, killing nine people, and injuring twenty more. The dog had saved the man's life.

∵ ❀ ∴

Incredibly enough, while Newfoundlands are often used in rescue attempts, they do not need to be trained to save a drowning person—their instincts tell them just what to do. The dog will first swim around the person. If the person is conscious, the dog will let them grab his tail or body. If the person is unconscious, the dog will grab them in his jaws and swim with them either to shore or to the nearest boat.

A Japanese girl named Hidomi became mute at age seven as a result of the abuse she had suffered by her parents. Her world was lonely and she felt that she couldn't trust anyone. She retreated into herself, feeling safe only in her isolation.

Because of the trauma, a court ordered her to be taken to live on her grandmother's farm where she would be away from her parents. On the farm were many animals. She began spending her days with them, petting and feeding them. She began to feel safe. They gave her unconditional love as she played with them. Eventually, she began to talk to the animals, regained her power of speech, and grew up to be healthy and happy. She is now studying to be a veterinarian.

∴ ❀ ∴

Animals that forage for food will often travel more slowly and over less difficult terrain while one of their group is suffering from an injury. This does not seem to fit into the model of survival of the fittest but is done purely out of consideration.

S ometimes "Prince Charming" is a dog. Kathy Quinn had been admitted to mental institutions thirty-six times in her life. Thought to be mentally retarded, she had been confined in straitjackets and suffered many of the abuses of institutional life. Nobody would have predicted that her life would turn into a success story.

One day, at a time when she was not institutionalized, she got a German shepherd. Walking with a powerful dog at her side, who would defend her against all others, gave her a confidence she could not find by herself.

However, she soon found herself in desperate trouble. Because of the dog's behavior problems, the authorities decided to have her dog destroyed. Her dog's plight forced her to do something that she would probably have never done for herself: Kathy undertook a training program with her dog to teach it to not behave aggressively toward other people. As she taught the dog gentle behavior, she had to approach people and situations that terrified her— an impossibility for her had she not been desperate to keep her dog.

As she rehabilitated her dog, Kathy rehabilitated herself to an extraordinary degree. She recovered from her illness so successfully that now she not only lives a normal life but has gone on to become an animal photographer and a famous trainer of dogs. Kathy Quinn's life has been completely transformed by love.

"The love for all living creatures is the most noble attribute of man."

—Charles Darwin

The dog of Henry F. Lewith, the author of the slogan "Be Kind to Animals" and promoter of Be Kind to Animals Week, refused to eat after Lewith died.

A toddler was playing in the backyard of his country home. His mother, who was in the kitchen, looked out the window to check on her baby and saw one of the most terrifying sights a mother can behold: A slavering, rabid dog was in the yard with her baby and was about to attack him.

The mother knew she had only seconds to reach her child, which might not be enough time. As she ran out the door toward her child she saw the family's goose attack the rabid dog. The goose bit the dog with its sharp beak and fought with all the strength and fury in its small body. Bloodied and in pain, the ferocious goose managed to divert the dog until the mother could grab her child, run inside, and call for help. The brave goose gave its life, but the baby was saved.

"By ethical conduct toward all creatures, we enter into a spiritual relationship with the universe."

—Albert Schweitzer

Tightropes in circuses must be just that—tight—otherwise the performer could lose balance and fall. Tightrope artists often personally tie or supervise the tightening of the ropes to make sure there are no mistakes.

During one act, however, the unthinkable happened and an artist was stranded on a slackening rope. Although the landing was just yards away, he couldn't get his footing and he knew he might die trying to reach it. The workers below tried to tighten the rope by tightening the turnbuckle, but only made matters worse. The rope got looser.

The crowd watched from below as the drama continued. Then Big Sue, an elephant, ran to the post by herself and with her strong trunk pulled on the anchor rope. The rope tightened, enabling the performer to make his way to the safety of the landing.

The audience gasped, realizing that the elephant had saved the performer's life. Incredibly relieved and safe at last, the tightrope walker looked down to signal his gratitude and saw Big Sue was still holding the rope taut. He tested it with his foot; it was as strong as any he had ever experienced. So in the best tradition of "the show must go on," he walked back out on the high wire, completed his act, and both he and Big Sue were wildly applauded at the end of the show.

Elephants may spend days and even weeks of courtship together before they actually mate. After they do finally mate, for about ten months the pair will be on a "honeymoon" during which time they will graze together, call each other when apart, and go for walks in the jungle.

"**A** Note on an Attempt to Induce a Schizophrenic-like State in Kittens" was the solemn scientific title given to a portion of a study done at the Veteran's Administration Hospital in Northport, New York. "Experimental kittens" were given 5,000 electric shocks when they were from one week to thirty-five days old.

The doctor who ran these experiments, Dr. Clarence Dennis, reported in the *Journal of Genetic Psychology* what he thought was an interesting side effect: "The behavior of the mother cat, although not part of the experiment, merits attention. When she eventually discovered that the experimental kitten was being given electric shocks during the feeding or whenever it was close to her body, she would do everything possible to thwart the experimenters with her claws, even trying to bite the electrodes off. She would run over to the kitten, trying to feed it or else comfort it as much as possible."

"I tremble for my species when I reflect that God is just."
—*Thomas Jefferson*

A young boy was leaning over the edge of a large tank of water at an aquatic show to get a better look. He stretched and craned his neck to see all the pretty fish. The tank was huge and he had never seen so many fish in his life. Suddenly he lost his balance and tumbled in. His mother screamed as her child sank toward the bottom of the gigantic tank.

People all around scurried in confusion. The boy tried to hold his breath but he couldn't swim. A bystander ran to the edge and was just about to jump in to save him when two large underwater shapes swept toward the boy. Two porpoises swam under the boy and carried him up to the surface, supporting him where he could breathe until the aquarium staff could get a hook and attach it to the child's clothes. He was then pulled to safety.

Wild dolphins have been known to circle an injured human to form a ring of protection from sharks. They will even drive their blunt snouts into the sides of the sharks so hard that their ribs break and the sharks cannot breathe.

Rae Anne Knitter, Ray Thomas, and Woodie, Rae Anne's dog, were walking along a trail in the woods. The scenery was beautiful and Ray wandered off to photograph a wonderful view from the top of a steep shale cliff.

Rae Anne was enjoying the day, when suddenly Woodie began pulling wildly on his leash. Sensing something was wrong, she dropped the leash and Woodie ran to the cliff edge and dove over. When Rae Anne reached the edge, she saw Ray lying eighty feet below in a stream, unconscious. The ground had given way beneath him as he stood on the edge of the cliff. But Woodie was with him, valiantly trying to keep Ray's face above water so he would not drown.

The devoted dog broke both hips diving over the cliff. But thanks to the dog's heroism, both Woodie and Ray lived! For his courageous rescue efforts, Woodie was named the 1980 Ken-L Ration Dog of the Year.

Although chimpanzees can't swim and are terrified of water, they have been known to leap into the water if there is a chance they can save another chimpanzee who has fallen in.

In England, a man once sold a heifer and her calf at the market. The mother was sold to Woolacott's farm and the calf to Sleeman's farm. The mother cow was taken to her new home, but in the night she broke away from the farm, through a hedge to a road. A mother's love, the two farmers were about to discover, is not a thing to be taken lightly.

The next morning, the cow was found at Sleeman's farm, seven miles away—a place she had never been before, happily suckling her hungry calf. Somehow she had been guided down strange roads and farmlands to find her baby. The mother was positively identified because she still had the auction numbers on her rump.

When Mr. Woolacott attempted to recover the heifer, his wife, who apparently better understood the importance of maternal instincts than he did, insisted he buy the calf, too.

The United States was the first country to enact animal rights legislation. In 1641, the Puritans included in their legal code: "No Man shall exercise any Tirrany or Cruelties towards any brute Creature which are usually kept for Man's use."

A South African publication documented an unusual fish story: the case of a deformed black moor called Blackie and another fish, a red randi named Big Red. Blackie had trouble swimming and for more than a year, Big Red had been swimming to his rescue daily.

The publication stated: "Big Red constantly watches over his sick buddy, gently picking him up on his broad back and swimming him around the tank. When feeding time approaches and their keeper sprinkles goldfish food on the surface, Big Red immediately picks up Blackie and swims him to the surface where both feed."

The real Andre, the seal portrayed in the movie Andre, *once stole one of the brass buttons from his little girl's coat and dropped it off a pier twenty feet into the ocean, just to watch it splash. It was a new coat and she was very proud of it but Andre could not resist the shiny buttons. Andre was scolded soundly for this act of theft by the little girl's father. The next day when the father went to the pier to visit Andre, the remorseful seal dropped the recovered button at his feet.*

The small orange kitten kept watch over a dying child. So steadfast was her vigil that the kitten refused to leave, even to eat. The child had cancer and was wasting away.

Even after the adults had gotten too sleepy to stay up any longer and had gone to bed, the little cat sat like a soft sentry near the child's head, calmly watching. Whenever the child awoke, she never felt alone in the darkness for she saw her kitten's beautiful gold eyes upon her. The child knew she never had to be alone or afraid.

Sometimes the kitten was sent outside by misunderstanding humans. She would then leap to the child's window and scratch on the glass, trying to get back to be at the child's bedside.

The child died. A little later, the kitten died of a similar type of cancer. Like a tiny angelic being, the kitten had helped her child move gently from this world to the next. But the people who had witnessed all this saw only a tiny orange kitten lying on a child's bed.

In ancient Japan, cats were held in such high regard that they were often kept on leashes; they were considered too precious to roam free. During the seventeenth century, when vermin overran the country and threatened the silk industry, the government declared all cats must be free and able to hunt. The country's economy was saved.

The *Oprah Winfrey Show* once featured an Irish setter named Lyric who saved her human's life by calling 911. The woman's machine for preventing sleep apnea (a disorder that can cause cessation of breathing and therefore death) had become unplugged in the night and she was beginning to suffocate and fall deeper into unconsciousness.

At first the Irish setter tried to rouse the woman by barking and pulling her bedclothes and jumping up on the bed but the woman did not awaken. The dog had been trained to get help and did just what she was supposed to do. She leapt up to the telephone where the woman had programmed a button for speed dialing 911. The operator traced the call and an ambulance responded.

> "Beneath this spot are deposited the remains of a being who is possessed of beauty without vanity, strength without insolence, courage without ferocity and all the virtues of man without his vices."
>
> —The epitaph on the grave of Lord Byron's dog

A young girl was out for her first horseback ride in Prospect Park, Brooklyn. The day was beautiful and it felt wonderful to be out in the fresh air and amongst the hills and trees, and yet, right in the middle of the city. Never having ridden before, she was just a little nervous but the horse seemed nice and gentle.

Suddenly, hoodlums decided to pelt her horse with stones and the panicked horse took off in a full gallop. She felt the horse bolt beneath her, its body tense and filled with fear and adrenaline. In her terror she dropped the horse's reins. Now there was nothing left for her to do but wrap her arms around the horse's neck and hold on, screaming.

A man on horseback witnessed the attack and took off after her. He galloped his horse next to hers, intending to pull her off. He knew it would be risky but he felt his only choice was to get her off the fear-maddened horse before she was killed. But then he noticed that the runaway horse was slowing down. He relaxed his grip on the girl for a moment and looked down to see what was happening.

His own horse had taken the runaway's reins in his mouth and was pulling it to a halt. The two horses came to a stop together.

My husband and I were walking down a street when we saw a female mallard get hit by a car. Her mate was present and obviously distressed. The motorist scooped up the body of the dead bird and took it away. Thinking that there was nothing we could do, we left the area.

Five hours later we happened to walk down the street again and saw that the mallard was still there, searching for his mate. This time we realized how serious the situation was to the little duck and we sat down as close as we could get to him. Speaking gently, we explained that his mate had been killed and that now he had to try to live his life without her. We told him that her death was final and that the reason he couldn't find her body was that the man whose car had killed her had taken her away. He looked at us as if he were actually listening and accepting. Then he sadly left the street and flew away in the direction of a nearby river.

"A man is truly ethical only when he obeys the compulsion to help all life he is able to assist and shrinks from injuring anything that lives."

   —*Albert Schweitzer*

Elsa Schneider of San Diego was blind and thought she should get a Seeing Eye dog. However, Elsa already had a cat named Rhubarb whom she loved, so she decided to try something no one would have believed could work: to train Rhubarb to be a Seeing Eye cat.

At first Rhubarb just jumped around and fought his leash; after all, he was an independent-minded cat, no mere dog, thank you very much! And yet, in time, he got used to the leash and seemed to sense her needs.

Now Rhubarb leads Elsa around the house and on small trips. When Elsa is outside and the telephone is ringing, Rhubarb comes and fetches her, guiding her into the house.

But Rhubarb retains a certain feline willfulness. He won't let Elsa talk on the phone too long. The cat puts his paws on her leg to signal that "you've talked enough."

W hen a wounded monkey crawled up on the doorstep of the west Bengal police station in New Delhi, India, little did anyone realize what an effect it was to have on the animal kingdom.

According to CNN reports, a school teacher had shot the monkey after he had wandered into the man's garden. The wounded animal hobbled to a nearby building that happened, by chance, to be the local police station and lay on the front steps. Neighbors found the monkey and took him to a veterinary clinic, but the doctor could not save him. The monkey died from the gunshot wound.

The dead monkey was then taken back to the police station. Shortly thereafter, at least fifty other furious primates gathered outside. In obvious disapproval at the murder of their friend by the villain, the monkeys began shrieking their protests. For hours the police station was surrounded by monkeys wailing their outrage and grief at the injustice of their friend being killed just for wanting to eat some fruit in a garden.

"The squirrel that you kill in jest, dies in earnest."
        —Henry David Thoreau

Mr. Bobcat Biscuit, a cat, was abandoned in an office building when his humans moved away. After some discussion, he was adopted by the office of social worker Kathy Kerr. Unconditional love proved to be Mr. Biscuit's forte and he turned out to be a huge help to Kerr and her colleagues, who work in child welfare services.

"A lot of times we have to bring the children in and talk to them. And they're pretty upset," said Kerr. But as soon as children arrive, Mr. Biscuit cozies up to them, purring with love and attention. "Usually we sit on the floor. The children can pet him or play with him. They don't have to look at us, and they usually start talking about things that have been going on in their lives."

Kerr also states that when she comes back from hearing a court case about child abuse, she sometimes feels like crying. But Mr. Biscuit sits on her lap and purrs until she feels she can do her job again.

"Teaching a child not to step on a caterpillar is as valuable to the child as it is to the caterpillar."

—Bradley Miller

J ames Wide worked at a railway station in South Africa. One horrible day he lost both of his legs in a train accident. Afterwards, he walked on wooden pegs or used a wheelchair.

James still needed to earn a living, and the only job he could do now at the railway was that of signalman. Painfully, he would drag himself to the railway switches to pull the levers that move the rails so that trains can switch tracks. The job was difficult but it kept him alive.

One day he realized that his pet baboon, Jack, could help him. Jack had been living with James for a while and had already learned to do the housework, the gardening, and pump water from the well.

James brought Jack to the railway station and taught him the work. Jack was good at the job. More and more, Jack learned to follow the guidance of James until he would just indicate to Jack which switches to pull and Jack would do the work.

Jack and James rode home each day on a trolley. For nine years, they were good friends as well as coworkers.

•.' 🐾 '.•

Ancient Egyptians kept baboons as workers. At least one appears in Egyptian art.

Dr. Arthur Peterson of Florida often watched the ducks on the lake on his property. He noticed that one of the male ducks was extraordinarily attentive to one of the females even though it wasn't mating season. Curious, he watched the pair until the male left for a few moments. Dr. Peterson captured the female and discovered that she was blind. Marveling as he understood the reason for the male's solicitude, he released her and the male duck immediately flew to her side and led her away.

A British miner was walking along a road when he saw two large rats walking slowly down the road together, each rat holding one end of a piece of straw in his mouth. Thinking to rid his countryside of one more pest, he clubbed one of rats to death. But to his amazement, the other rat didn't run. The man stooped down to study the rat more closely. It was blind.

Ways to Return
the Kindness

❦ *Avoid feeding the birds around your home during autumn. It spoils some of them and they don't want to migrate south in the winter and then they freeze. But do feed them in the middle of winter so that those who had to stay behind do not starve. Make sure the water in your birdbath is replaced frequently and unfrozen.*

❦ *When you clean out the lint in your laundry dryer, if you place it on a branch of a tree, birds can use it to line their nests. The soft, moldable material is comfortable for baby birds.*

❦ *Consider volunteering at your local SPCA. You can play with and pet the cats, and the SPCA has dog walker programs. There are few things you can do for the animals that are as wonderful as taking time to walk dogs at the shelter. They are often lost and abandoned pets who are lonely and confused about the turn their lives have taken. If you live in a building that doesn't allow dogs or if you have a child that wants a dog but can't have one, then by all means do both yourself and the animals a favor and join a dog walker program.*

❧ *Don't leave cloth bedding in a dog house in the winter. If it gets wet, it will freeze.*

❧ *Never buy animals at dimestores or large department stores that leave their animals to starve over holidays and weekends.*

❧ *Speak out against cruelty wherever you see it.*

❧ *The most important kind act of all: There are fifteen dogs and forty-five cats born for every man, woman, and child in the United States at present. Homes can never be found for them when there are so many. So please—SPAY and NEUTER your companion animals.*

From 1880 to the early 1900s, a dolphin named Pelorus Jack guided ships through the straits off New Zealand. As soon as he heard a ship coming, Pelorus Jack would rush to it and leap into the air. The excited crew, who from many years experience knew Pelorus Jack, would shout when they saw him, knowing that they would be safely led through the dangerous waters. His guidance was considered so valuable (for only a dolphin could swim underwater, detect the swiftly moving currents and see the sandbars) that a government edict was issued protecting him.

One day a drunken man aboard a ship called the *Penguin* shot Pelorus Jack, and the dolphin was presumed dead. Several months later he returned and, to the sailors' relief, continued to aid ships entering the straits. But whenever the *Penguin* came by, Pelorus Jack would not help it. How he knew which ship the *Penguin* was no one could say; perhaps he could tell that particular ship's engine noise.

Pelorus Jack continued to guide ships safely through the straits for many years—all but the *Penguin,* which sank in 1909. All seventy-five people aboard were drowned. No wonder so many sailors believe it is bad luck to hurt a dolphin.

Sometimes altruistic behavior becomes so prevalent in a species that it becomes part of the instinctual nature of that species.

When a pair of red-cockaded woodpeckers nest high up in an old pine tree, they do not have to try to protect their nest alone. Two to four adolescent birds help them. These "babysitter birds" aid in building the nest, defending the territory, sitting on the eggs until they hatch, and feeding the young. Most of the time the young helper birds are the offspring of the pair from a previous season, but not always.

Red-cockaded woodpeckers are an endangered species.

Truck driver Martin Dacar was in his garage attempting to change a 2,680-pound tire. As he shifted it into position, the weight proved too much for him and it began to fall onto him. Martin dove away but the tire fell on top of him, pinning his foot against his leg.

He assessed his situation. It was night, he was alone in a deserted warehouse, the temperature was 25° below zero, he had broken bones, and he was losing blood. He didn't think he would make it until morning. The cold was already beginning to make his body go numb. Martin reached for some tools on the warehouse floor to wedge them under the tire, but he couldn't raise it high enough. He tried lighting a fire to keep warm and to attract attention, but he couldn't get it going.

Then he heard dogs barking in the distance. If he could get their attention, perhaps someone might come to see what the fuss was about. Martin called to the dogs. They barked back. He called again. They returned his call. Finally a woman named Annaliese Schmidt sent her husband, Harvey, to see what the usually quiet dogs were so upset about.

Martin was saved and Mr. Schmidt later told him that one of the dogs, Buddy, had himself been rescued by Mr. Schmidt years before. Still not much more than a puppy, Mr. Schmidt had found him, hungry and cold, under a bus in the snow and had

taken him in. Martin became fast friends with the
Schmidts and their dogs, Buddy and Champ, often
bringing his two saviors dog biscuits.

*"I think they're the same thing. It's a feeling for
life—all life."*

— *Princess Grace of Monaco, when asked about
the difference between charitable causes for
people and those for animals*

Diagnosed with terminal cancer, a woman deep in depression happened to find a tiny, starving, sick kitten under her car. She cleaned her, fed her, and gave the cat medicine until she was well. The kitten became the bright spot in her life.

Then the kitten was diagnosed with feline leukemia. The woman, feeling she couldn't handle any more heartache, began to ignore the kitten and retreat emotionally from it.

But the kitten didn't ignore her. Not realizing it had leukemia, it continued bouncing around just as happily as ever, a constant reminder of simple delight in being alive. The woman realized that while she had been studying mind-over-matter methods of healing for herself, trying to accept the idea that the body could heal any disease, she didn't quite believe it. But her cat did.

The woman resolved to reinvigorate her program of alternative healing along with the traditional methods. And she incorporated her feline co-patient into the regimen. If the woman got a massage, the cat got a massage. If the woman had healing crystals in her bedroom, the cat had crystals in its water bowl. The kitten had made her realize she had to love now, whatever the future might hold, and have faith. Seven years later both are still loving life and the cat's last three leukemia check-ups proved negative.

When pets were introduced to one ward at the Ohio State Hospital for the Criminally Insane, levels of violence dropped drastically. Prisoners made fewer suicide attempts and their medication requirements decreased. The prisoners' interaction and communication with their doctors was significantly improved. However, acts of violence in other wards increased—because prisoners in those wards wanted pets as well. Finally, the prisoners with pets shared the offspring of their animals with the other wards.

It was one of the saddest sights I have ever seen: a small brown bird was trying to dive under the wheels of moving cars as they drove down a city street.

I thought, "What are you doing, you crazy little bird?," but then I saw the squashed body of his mate flattened against the asphalt and I understood. The little bird continued to fly as close as he could get to his dead mate, sometimes even passing under the cars, narrowly missing being crushed by their wheels. Apparently trying to rouse his mate and give her encouragement to fly away again, the bird made attempt after attempt. Helplessly I watched as he risked his own life in its desperation. The cars sped past, unheeding and unaware of the little drama that was unfolding.

This tiny bird seemed to me as valiant in his devotion, courage, and love for his beloved as any hero in literature.

"If a man aspires towards a righteous life, his first act is abstinence of injury to animals."
—Leo Tolstoy

A Seeing Eye dog must sometimes intuit danger where it might not be obvious to another dog. Nemo, a German shepherd, was walking with his blind mistress, Mary, one night after a storm. They always enjoyed their walks together but Nemo abruptly pulled Mary on a quick detour. Mary fully trusted her dog's instincts and, although the route was unfamiliar, she obeyed her dog's direction.

She might never have known how close to danger she was. She later learned that the storm had broken power lines and a live wire was dangling on the path around which Nemo had so carefully steered her.

∴ 🐾 ∴

Seeing Eye dogs have to be keenly aware of the area around their bodies. They must also take into account their person's additional height. Even if the way is clear around a blind person's feet or where a cane could tap, a Seeing Eye dog will carefully steer his person around a lofty overhang that the person's head would otherwise have struck.

When Ann Landers wrote a column disparaging birds, one of her readers wrote in about the following heartbreaking incident. The reader used to let his two parakeets fly free around his home. Usually they would both fly to greet him when he came home from work.

This time, though, one of the birds was screeching wildly as he entered the house. The other was lying dead on the living room floor while his mate was flying in circles around his lifeless body. The man tried to catch her but she refused to come to him as she had always done in the past. Instead, she flew into the bathroom and drowned herself in the toilet.

"He would daily throw out crumbs for the sparrows in the neighborhood. He noticed that one sparrow was injured, so that it had difficulty getting about. But he was interested to discover that the other sparrows, apparently by mutual agreement, would leave the crumbs that lay nearest their crippled comrade, so that he could get his share, undisturbed."

—*Albert Schweitzer, of a friend's experience*

Bo, a Labrador retriever, and Duchess, a puppy, were rafting down the Colorado river with their humans, Laurie and Rob Roberts.

Suddenly a wave toppled the raft, pinning Laurie and Bo underneath the water. Bo struggled free, but realizing that Laurie was still underneath, he dove back under the raft for her and pulled her out by her hair. Gasping for air, Laurie made it to the surface. She had grabbed Bo's tail and the strong dog pulled her through the dangerous currents to the shore.

> *"I care not for a man's religion whose dog or cat are not the better for it."*
>
> —*Abraham Lincoln*

Mary, an avian member of the National Pigeon Service during World War I, was missing and it was presumed that she had been killed. She had never been late before and those who knew her were sure that only death could stop her.

And it almost had. She had been attacked by a hawk and her neck and right breast were ripped open. Although it took her four painful days to do it, she got through the German lines.

During her next mission, Mary was absent for three weeks. She must have braved hunger, pain, and predators while she was too weak to fly and had lain on the ground, trying to nurse her wounds. This time she had three pellets in her body and part of a wing shot off.

Later, a 1000-pound German bomb exploded just outside her loft, killing most of the people in the area but Mary, somehow, survived.

The need for brave carrier pigeons was so great that Mary was again returned to active duty. On her last mission, the tiny survivor was discovered with a wound that had opened up the side of her head and neck as she tried to bear her message to the allies. She was given stitches and nursed back to health.

"If [man] is not to stifle human feelings, he must practice kindness toward animals, for he who is cruel to animals becomes hard also in his dealings with men. We can judge the heart of a man by his treatment of animals."

—Immanuel Kant

The passenger pigeon was once so plentiful that historians record that a flock could blacken the skies for days. Over two billion of them were sighted in one flock. Seven and one-half million were killed in one hunting expedition alone. Deforestation and wanton hunting decimated the flocks. When Martha, the last carrier pigeon, died on September 1, 1914, in a Cincinnati zoo, the carrier pigeon became extinct.

Bill, a resident of Oxford, England, had suffered from clinical depression for more than ten years. Horace Dobbs, who runs International Dolphin Watch, decided to take Bill out in a boat off the Pembrokeshire coast to see if watching dolphins would have a positive effect on the man. A dolphin swam right up to Bill, although there were twenty other people in the boat.

Communicating with this creature began to lift his depression.

"I felt wanted for the first time," Bill said. "There were no questions asked." The dolphin stayed with Bill. "The message I received was, 'I need you and you need me. Let's share our lonely worlds together.'"

Bill was so impressed with the experience that he decided to swim with another dolphin off the coast of Ireland. Bill now returns to swim with that dolphin every year.

B o is a dog who works with Brad Gabrielson, a cerebral palsy victim in Jamestown, North Dakota. One day, Brad fell out of his wheelchair. Unable to get enough control over his body to rise, he was terrified that he would have to spend a whole day painfully stranded on the floor. He could not even get enough control to have Bo help him back up. He felt utterly helpless, and knew it would be many hours before someone else would be home.

But Bo knew what to do. Realizing that Brad needed human assistance, Bo pulled a string tied to a lever that unlocked the front door and ran through the streets to a neighbor's house. Bo roused the neighbor, very gently taking the startled neighbor's fingers in his mouth, and guided him to Brad's home. The neighbor was glad to help and Brad was saved from a frightening day of waiting alone until help arrived.

Some dogs who work with epileptics can sense when their person is about to have a seizure and will gently guide them to bed. No one knows how they can tell this.

One day a naturalist gave a chicken hen twenty-one guinea fowl eggs just to see what she would do. The small eggs were very different from chicken eggs but she sat on them anyway and the naturalist just assumed that she was too dumb to notice the difference. Eventually, the eggs hatched and little guinea fowls came out, but the mother hen was unperturbed. Again the naturalist thought she just didn't have the intelligence to perceive that these were not baby chickens.

Then she did something that absolutely shocked him. She did not lead the little chicks to the feed that chickens need. Rather, the good adoptive mother led them to some bushes and scratched at ants' nests for pupae. Now chickens don't eat such things and they don't scratch for them but guinea fowl most certainly do! The little ones ate until they were full.

Another naturalist gave a hen some duck eggs to nurture. After they hatched she did what no chicken in the world would ever do with her own chicks. She led them to water and cajoled them to swim.

When a fire broke out in their house, the family's fox terrier barked and frantically tried to alert her sleeping humans. The entire family was asleep but the little dog barked and pulled at the covers until they realized that smoke was filling the house. As flames leapt through the home and the horrible heat increased, a fireman rescued the three children. The family was safe.

Then, and only then, did the valiant little dog rush back into the blaze and save her own puppies.

> "The fidelity of a dog is a precious gift demanding no less binding moral responsibility than the friendship of a human being."
>
> —Konrad Lorenz

B uddy, America's first and most famous Seeing Eye dog, was taught not only to obey the commands of his human but also to refuse when there was a danger of which the person might not be aware.

Once when Morris Frank, the blind man Buddy guided, was attempting to get on an elevator in a hotel, Buddy refused to move forward no matter how many times the command was given. Exasperated with his disobedient dog, Frank stepped forward anyway whereupon Buddy threw his body against him and refused to let him move. Then Frank heard a maid behind him scream: "Don't go near that elevator! The car isn't there and it's nothing but a big hole."

A dinner guest at an event Morris Frank attended turned out to be the daughter of a railway executive on the very railway line that Frank was trying to persuade to allow Seeing Eye dogs aboard with their blind persons. At that time, even Seeing Eye dogs were relegated to the baggage cars. Without knowing who she was, Frank told her about his favorite subject—Buddy.

The dog, he told the friendly guest, had turned Frank's life around. He went from being a helpless invalid, who could not even go out for a haircut by himself to a self-sufficient young man with great enthusiasm for life. He traveled extensively, lecturing about the new freedom that Seeing Eye dogs could bring to blind people.

Frank later found out that after their conversation his dinner acquaintance had discussed the matter with her father. She had even refused to eat Christmas dinner with her family "until those dogs are treated like the human beings they really are." Her father relented and the dogs were subsequently allowed on trains.

> "Man is the only slave. And he is the only animal who enslaves. He has always been a slave in one form or another, and has always held other slaves in bondage under him in one way or another. In our day, he is always some man's slave for wages, and does that man's work; and this slave has other slaves under him for minor wages, and they do his work. The higher animals are the only ones who exclusively do their own work and provide their own living."
>
> —Mark Twain

Dolphins understand the difference between play and a serious situation. Once, the crew of the boat *Aquanaut* had to give up its plan to practice lifesaving techniques in the ocean because a playful, happy dolphin kept interrupting the activities.

Later in the day, when a member of the boating party got into serious trouble, the dolphin gently supported the man on the surface and helped a crew member tow him to the diving ladder. But that was not the end of the dolphin's concern. The dolphin swam alongside the ship and watched quietly until he could see that the man had recovered.

"Perhaps, in some way, I owe my gold medals to the dolphins. In their trusting and playful way, they taught me the subtleties of swimming technique."

—*Olympic gold medalist Matt Biondi, who swims with the dolphins*

The destruction of Pompeii on August 24, 79 AD by volcanic eruption was so sudden that people had only a moment's warning and could not escape. But of the 2000 skeletons found almost nineteen centuries later, none told a story so poignant as that of the form of a dog stretched protectively over that of a child. Hot ash had rained down upon the city, entombing all. Uncovering the ancient dog and child bodies, the excavators were moved by the silent tableau of faithfulness to the last breath.

But they were even more amazed when they deciphered the writing on the dog's collar. The finely etched inscription told how the dog had saved the child's father, Severinus, three times— once from drowning, once from thieves during an ambush, and once from a wolf. The dog's dying action was to give his body in the brave but futile protection of the child he loved.

In the third century BC, the Emperor Asoka of India declared that animals must be treated with reverence throughout his empire.

"Medic!" the wounded soldiers called out from the mud where they lay, trapped in their foxholes. But shells were exploding and shrapnel cut through the air all around them; no medics could get through the field to their foxholes alive. Fires fed by phosphorus bombs lit up the night with unholy brightness as the Japanese fighter planes strafed the ground.

"Call Chink!" a medic yelled as he used adhesive tape to attach packets of medicine, morphine, and bandages to the German shepherd's thick fur. The wounded soldiers got the idea and began to call to Chink. The company's large mascot dog crawled through the explosions and machine-gun fire and into the foxhole of each wounded soldier. The injured soldiers tore off what they needed and the dog crawled back through the gunfire to the next foxhole where another desperate soldier awaited her.

Chink continued to crawl on her belly from foxhole to foxhole. She was very aware of the danger; twice she was hit with shrapnel. But still she struggled on and the soldiers got the medicine they needed. The next day Chink was treated for her wounds, the shrapnel was removed, and she recovered.

A small, female dolphin was badly injured by a hook used in her capture. She was so wounded and in shock that the researchers who had captured her feared she would die. She was unable to resurface for air as dolphins must if they are to breathe. In desperation, the researchers tied large glass jars to her body which made her float so that she could breathe, but her despondency was so severe that they were still losing her.

Finally it was decided to introduce a male dolphin into her tank. She responded immediately and with great joy, making greeting sounds to her new companion. Happy again, she even tried to swim. The researchers then removed the glass jars and the male swam with her, nudging and lifting her to the surface whenever she needed air. Again and again he would assist her with love and patience. Because she was dependent upon him for the very air she needed to breathe, his devotion had to be flawless.

At last she began to recover and the two dolphins would swim, play and rub together affectionately for hours. For two beautiful months, they enjoyed their friendship together. Then the female suddenly died from her wounds. The inner trauma had been too great.

The male was utterly despondent. As she died, he circled round and round her, crying out calls of distress. Afterwards, he refused to eat and swam in circles calling for her for three days until he, too, died.

> "When the male and female beavers who mate for life are forced apart, they cry as we would."
>
> —V. V. Krinitsky, Russian Delegation on Environmentalism

John Lilly, a scientist studying dolphin intelligence, eventually freed his dolphin subjects, releasing them back to the ocean. He explained: "I felt I had no right to hold dolphins in concentration camps for my convenience."

The kitten was absolutely terrified. Several young Liverpool hoodlums were throwing rocks at her, trying to stone her to death. With each stone, the little kitten hurt so badly it seemed she could never survive. Each time she tried to escape, the boys recaptured her and hurt her again and again.

At last one cruel boy grabbed her and carried her toward a dirty pond to drown. A dog was wandering by and saw the boys begin to drown the kitten in the pond. Although cats and dogs are supposed to be enemies, apparently the dog could not put up with such cruelty.

The outraged dog leaped on each of the boys in turn and sent them running. Then he rescued the terrified kitten from the pond and, licking her wounds, took her to his home where he shared his own food with the poor creature.

Miraculously, the kitten recovered and the two grew to be fast friends. The unusual cat-dog bonding became a famous attraction at the Talbot Inn in Liverpool, where they lived out the rest of their lives together.

Tina, an elephant in the Central Park Zoo, trusted only one handler, a man named Robert Brockell. When Robert became sick with leukemia, Tina refused to go inside to her quarters because her "main man" had not spoken the direction.

As winter approached, this became a serious concern. No one else could persuade her. She responded to an attempt at forcing her inside with anger, injuring the new would-be handler. Desperate zoo officials finally made a recording of Robert at the hospital, so she could hear his voice ordering her to go inside. But Tina refused to obey a mere recording.

Finally Robert volunteered to be taken to the zoo in an ambulance. He was delivered to the elephant's side on a gurney. "Go inside, Tina," he said.

She obeyed.

Christine Harrison was visiting her parents in Barnsley, England with her Chihuahua, Percy. Percy panicked and ran into the street where he was struck by a car. Saddened by the loss of her beloved dog, Christine wrapped him in a heavy paper sack and buried him.

Soon, Christine's parents' dog, a terrier named Mick, began acting strangely. He barked and ran to the spot where Percy was buried and dug up the new grave. Then he dragged the sack into the house, shook Percy's lifeless body out and began licking the little dog.

Christine and her parents finally noticed a faint heartbeat and rushed the Chihuahua to the vet's. There the animal recovered. The veterinarian surmised that Percy had enough air in his sack to survive being buried alive but was in shock and that Mick's licking had nudged the Chihuahua's circulation just enough to revive him.

What surprised Christine and her parent's more than anything else about this unusual rescue was that Mick, the parents' terrier, had always hated the little dog.

J ames Wentworth Day, the wildlife writer, related this incident that was described to him by Commander David Blunt, who held the title of Cultivation Protector of Tanganyika. An African woman had placed her baby in the shade of a tree while she worked. An elephant herd strolled by and saw the baby. Several of the elephants pulled leafy branches from the tree and covered the sleeping babe with them. Flies can be a problem in this part of Africa and the branches protected the infant from the flies. The elephants were so gentle and quiet about this that they did not even wake the baby.

The elephants then departed.

I will remember what I was
I am sick of rope and chain.
I will remember my old strength
And all of my forest affairs.
I will not sell my back to man
For a bundle of sugar-cane.
I will go out to my own kind
And the wood-folk in their lairs.
I will go out until the day.
Until the morning break,
Out to the winds' untainted kiss,
The water's clean caress.
I will forget my ankle-ring
And snap my picket-stake.
I will revisit my lost loves
And playmates, masterless.

     —Rudyard Kipling, "The Captive's Dream"

I n June 1971, Yvonne Vladislavich was sailing on
a yacht in the middle of the Indian Ocean when
suddenly the craft exploded. She was thrown clear
but the vessel sank and she was left completely
stranded. Far from the shipping lanes, there was no
hope of rescue.

Terrified, she treaded water awaiting certain
death. Then she saw three dolphins approach her.
To her astonishment, one of them swam underneath
her and buoyed her up with his own large body.
Gratefully, she held on to the dolphin's sleek,
smooth body. The other two dolphins swam in
circles around her to protect her from sharks.

The dolphins carried and protected her through
the warm waters for many hours until they arrived
at a marker-buoy floating at sea. They left her on
the buoy from which she was soon picked up by a
passing ship.

It was calculated from the position of the buoy
and the position of her yacht when it exploded,
that the dolphins had carried her and kept her alive
through two hundred miles of dangerous seas.

Bobby, a homeless Scottish terrier, scrounged food wherever he could find it. A scruffy, unattractive little dog, he had learned that he could expect only cruelty from people.

One old man, an elderly shepherd named Auld Jock, was kind to the little dog and bought him some scraps at a local restaurant. It was just a little bit of love but it meant so much to the lonely dog. When the old man died, Bobby set up a vigil at the man's grave site that lasted fourteen years.

At first, the grave diggers ordered him away, kicking him and throwing rocks at him, but nothing could persuade the dog to go. Through cold winters and harsh summers, he stayed at the grave site, leaving only to beg or steal food from the town. Any food he got (which was usually just a little bun) he would bring back to the grave and eat there. In his first year of vigil, Bobby slept under a tombstone. The next winter the townspeople built a little shelter just for him. Finally, the villagers were so touched by the little dog's loyalty that when Bobby died they erected a statue in the town square in his honor. It still stands in Edinburgh, Scotland, at Greyfriar Square. But Bobby's body was buried next to the body of Auld Jock, his friend.

They say that what goes around comes around, and that saying is certainly proven true in this story. Michael Zezima sometimes fed a stray dog who lived in his neighborhood. This simple act of kindness was later to save his life.

During the blizzard of 1978, the New York area was pummeled by a severe snow storm. Michael was walking home and came upon a woman who couldn't make it to her own house because of the waist-high drifts. He helped her to the home of one of her friends and then took off towards his own home which was only two blocks away. But the snow and his exhaustion from helping the woman proved too much for him.

Michael collapsed in the street.

Cold and alone, his body stiff and numb, Michael thought he would die. Then the dog that Michael had been feeding chanced upon him. Seeing his friend's desperate condition, the dog lay down on top of Michael and kept him warm. As he cuddled Michael he barked loudly for help. For three hours the two lay locked in embrace, their bodies creating the only warmth they had to keep them alive. Finally someone heard the dog's persistent barking. Michael was rescued and taken to the hospital. The dog refused to leave Michael's side and the doctors decided to allow him to stay at the hospital that day.

Michael recovered and adopted the dog.

In a children's hospital, as part of a program to bring animals into hospitals, a little deaf girl was shown a two-year-old chimpanzee. The child was so sick she had to be wheeled in on a gurney. Suddenly the girl started to cry. As it happened, the chimpanzee had been trained in a sign language. To everyone's astonishment, he began signing to her. The little girl forgot her sadness and the two became friends.

A steamer ran aground off the coast of Newfoundland. The waters were rough and the ship was coming apart at the seams. Panic swept over the passengers and it seemed they would all drown. The people on the shore could only watch helplessly because the waters seemed too rough to even attempt a rescue.

But a man on shore had a Newfoundland dog and he attached a line to the dog's neck. The great Newfoundland dove into the icy, turbulent waters and, following the directions of the man, swam to the ship. A lifeline was established and a conveyor device was sent along the line to the ship.

One by one the ninety-two passengers aboard got into the conveyor and were pulled to safety. One time the conveyor reached shore with a mailbag inside. It contained a baby. The conveyor went out again and again, as the ship broke apart, until there was only one man left aboard. The rescuers were surprised when they pulled the conveyor in to find, not a man in it, but the Newfoundland dog. The last man had decided to take his chances on the crumbling ship and make sure the hero dog was saved. The conveyor went out one last time and all ninety-two passengers were saved. The dog was later awarded a medal of honor.

Even animals that have come to symbolize ruthlessness and ferocity can be kind. Gordon Haber, an expert on wolf behavior, was watching a wolf whose shoulder had been shattered from a kick by a caribou. The wolf wandered off to be alone and perhaps to die, but every day another wolf devotedly brought meat to his crippled friend. The injured wolf was fed this way until he recovered.

> *"The worst sin towards our fellow creatures is not to hate them, but to be indifferent to them. That's the essence of inhumanity."*
>
> —George Bernard Shaw

The Soviet news agency Tass reported that the sailors on a fishing vessel saw a ring of killer whales about to devour a sea lion that was bleating for help. A group of dolphins rushed to the rescue and surrounded the sea lion, thereby protecting it.

When the dolphins left, however, the whales returned and the sea lion again called for help. The dolphins returned a second time, and, in a thrilling gesture, leaped over the killer whales, surrounded the sea lion and stayed until the killer whales were gone.

If a group of sperm whales is sighted by a whaling ship, the whalers have only to kill or injure one to get them all. The others always surround their injured friend to help him. Then the hunters can pick them off one by one.

A captive orca likes to look at a picture book called Killer Whales *which has pictures of other orcas in it. John Ford, the scientist who cares for her, says she doesn't care to look at books with other animals in them, only orcas.*

Anthony, a mentally handicapped boy, was swimming in a lake near Houston, Texas, when he realized he had gone too far from shore. He panicked and began to sink. A pet pig, Priscilla, jumped into the water, swam out to the boy and, once the boy had grasped the pig's leash, towed him back to shore.

The pig became a local celebrity.

When bacon sales dropped substantially a few years ago, marketing research was done to find out why. It turned out that the movie Babe *had just come out and kids across the nation were refusing to eat bacon.*

"Dogs look up to us. Cats look down on us. But a pig will look a man in the eye and see his equal."
—Winston Churchill

A troop of baboons were walking to their favorite water hole when suddenly they were surrounded on all sides. They had walked into a trap, which shut behind them, permitting no escape. They started screaming and shaking the sides of the trap but could not break through.

As the hunters moved in to collect their quarry, they encountered hordes of free baboons who raced to aid their screaming caged comrades. The baboons were so outraged that they chased the hunters away even though the hunters had guns. Then the free baboons attacked the trap with all their fury. Using their combined strength, they dismantled the trap and set their brethren free.

Sammi, a very successful hunter cat, always left one or two field mice outside Susan's door each morning. These were the mighty hunter's gifts of love.

When Susan moved to a small apartment in a more urban environment, there were no mice to be found and only a few trees to climb. However, the resourceful Sammi was not to be discouraged. He took to leaving two pine cones outside the door and continued to do so until he died.

Susan says in the book *Animals as Teachers and Healers*: "My little cat showed me that no matter what the situation we can find a way to make the best of it. To this day, when things are hard to deal with, I see her at the door with those pine cones."

"Reverence for life comprises the whole ethic of love in its deepest and highest sense. It is the source of constant renewal for the individual and for mankind."

—Albert Schweitzer

Near Auckland, New Zealand, fifty pilot whales became stranded in the shallow waters of a harbor. Realizing that the whales would become beached and die, the local men tried everything to chase them out. Nothing worked.

Finally, a local official got the idea of using speedboats to guide a passing group of dolphins into the harbor. Once there, the dolphins seemed to immediately understand the situation and guided the pilot whales out.

⠒ 🐾 ⠒

A researcher, trying to paint marks on a young rhinoceros for tracking purposes, discovered that its fearful cries brought not only its mother but many other unrelated rhinos who rushed over to defend it.

A researcher studying wild capuchin monkeys in Venezuela noticed a baby born with partially paralyzed legs. Because it could climb but not jump from tree to tree, it was carried more than others for its age.

In 1975, a woman was shipwrecked in the waters off the coast of Manila. Utterly alone and far from land, she began to pray.

Then she saw a giant turtle swim toward her. Wildly, she clutched its shell and the turtle began to swim away with her. For two days the turtle swam and towed her toward land, refusing to dive— which sea turtles need to do to hunt for food. The turtle gave up eating in its commitment to save the stranded woman's life.

When she was finally picked up by a ship, the crew thought she had been clinging to an empty oil drum—until the "drum" swam in circles and, seeing that she was all right, disappeared beneath the waves.

Barry was a beautiful, strong Saint Bernard dog who lived from 1800 to 1814. During his lifetime, Barry rescued forty people from the snows of the Alps.

He had an incredible ability to predict where avalanches were going to occur. Before helicopters were in use and could spot people in the snow, Barry was a snowbound traveler's only hope. Every morning Barry would, of his own volition, race up the mountains and look for fallen wayfarers. He led the monks of the Hospice of Saint Bernard to many travelers who were buried in the snow. After his death, the monks continued to name their lead dog Barry in his honor.

Jaco, a parrot in Salzburg, could not only speak but seemed to understand grammar. Whenever his person left, Jaco would say "God be with you." But when several people were departing, Jaco would change it to "God be with all of you."

One of the first examples of the effects of animals in institutional therapy happened by accident. In the early 1970s, Dr. S. Corson was a professor at Ohio State University, working in both psychiatry and biophysics.

One day, a young patient at the mental hospital heard the barking of the dogs upon whom Corson was performing medical experiments and asked if he could have one. The boy got his dog, as did other patients who also began requesting them, and the results were amazing.

Patients who failed to respond to conventional treatments, such as drugs and electric shock, suddenly began to get well. Patients who had refused to speak or leave their beds got so much better that some were even discharged from the hospital. Corson described his results in a paper called "Current Psychiatric Therapies" which was very well received by the rest of the medical community. The natural attraction between living creatures to love and nurture each other proved to be a most potent form of healing.

In one study of convicts, those who were allowed to keep cats were all rehabilitated.

I imagine you are a handicapped child and all your life you struggled to walk and do all the things other kids do so easily. You felt insecure and inept and you couldn't accomplish even the easiest of tasks. Now imagine that one day you go to a special camp and you are set upon a horse which lifts you high above the ground, and suddenly, you have strong legs under you. And you also know that with a special harness and with training you will finally be able to ride that horse and run like the wind.

This is the experience many children have at the over 200 riding centers for the handicapped persons in America. These programs bring joy and self-esteem to untold numbers of youngsters. Here, at last, they can feel the freedom of going where they choose without anyone else having to help them. For many children this experience is the beginning of a self-confidence and a re-evaluation of what is possible. The hope these horses offer lasts a lifetime.

"To help life reach full development, the good person is a friend of all living things."
—Albert Schweitzer

Malakoff, a very large Newfoundland dog, was the watchdog for a Parisian jeweler. One of the jeweler's apprentices, a man named Jacques, hated Malakoff who, perhaps, sensed something about the man that he did not trust.

Jacques resolved to kill the dog.

With a few other cohorts, Jacques led the great dog to the river Seine, tied a stone around his neck, and threw him into the fast-moving water. Malakoff fought for his life, swimming and struggling for the shore. He swam so powerfully that even with the stone he managed to make it almost to the shore. Then Malakoff looked behind him and realized that his attacker, Jacques, had fallen into the water, too, and was drowning. The man gulped for air as he thrashed in the water but, not knowing how to swim, he panicked and started to go down.

Seeing this, Malakoff turned and swam back toward Jacques. Despite the heavy weight around his neck, Malakoff swam, panting and straining, to where his would-be assassin struggled. In desperation, the man grabbed Malakoff's fur. By now too weak to pull the man to shore in the strong cross-currents, Malakoff struggled with all his might just to tread water with both the stone weight and the panicky man. The dog held Jacques afloat until others could rescue him.

Once man and dog were both safely on shore, the remorseful apprentice threw his arms around the great Newfoundland and wept as he begged the dog's forgiveness.

The story of the heroic dog spread throughout Paris. Malakoff became such a symbol of valor that when he died, nearly every apprentice in Paris followed his funeral procession.

"Our task must be to free ourselves by widening our circle of compassion to embrace all living creatures and the whole of nature in its beauty."

> —Albert Einstein

"The animals of the world exist for their own reasons. They were not made for humans any more than black people were made for whites, or women created for men."

> —Alice Walker

Senseless Acts
of Beauty

Bower birds decorate the walls of their elaborate nests with flowers, ribbons if they can find them, and other pretty things. The flowers are chosen for their beauty, and not just as nesting material, because as the flowers wilt, the birds replace them with fresh flowers.

Drawing by Siri the Elephant

Anne Herbert's original quotation inspiring the Random Acts of Kindness series was "practice random kindness and senseless acts of beauty." Many people like me may be thinking, "If animals can do such beautiful and selfless acts of kindness, then what is it that does separate us from the animals?"

Perhaps it is art. Surely only humans can create art. That is what I thought, too, before I started doing the research for this book. It turns out that even here the human race gets a comeuppance. David Gucwa was working as a zookeeper at the Burnet Park Zoo in Syracuse, New York, when he noticed the scratches made on the cage floor with a stick by an elephant named Siri. The scratches were complex, actually rather lovely. Curious, he gave her a pencil and some paper just to see what she might do.

At first Siri just chewed on the pencil or used it to scratch herself. But after a while, she discovered its true purpose and started making drawings on the paper. The drawings did not seem to be random scrawls, but showed an interesting form and grace.

The result was a series of pictures that struck David as so interesting that he took them to well-known artist Jerome Witkin without mentioning who the artist was. Jerome was busy completing paintings for an exhibition entitled "Jerome Witkin:

A Ten-Year Retrospective" but took time out to view the work of the mystery artist.

Jerome was delighted with the works. "These drawings are very lyrical, very, very beautiful," he stated. "They are so positive and affirmative and tense, the energy is so compact and controlled, it's just incredible." When he was told the nature of the 8,400-pound artist, Witkin was not offended as David had feared, but stated that he was even more impressed. "Our egos as human beings have prevented us for too long from watching for the possibility of artistic expression in other beings," he said.

David continued to work with Siri as she produced an amazing volume of work over several years. To compare her output with those of other elephants, he contacted the Shrine Circus when they came to Syracuse. He asked the elephant handler if he had ever seen any of his elephants pick up a pebble or stick and doodle on the ground. "Sure, they do it all the time," the man replied, "So what?"

David contacted the curator at the Washington Park Zoo in Portland, Oregon, and asked him if he had ever seen anything similar. "All our elephants draw," the curator said, and added that the zoo staff didn't think much about it.

David later sent some samples of Siri's work to the world-renowned artist Willem de

Kooning. "That's a damned talented elephant," was de Kooning's response, while Mrs. de Kooning expressed an interest in following the elephant's career.

Elephants in other zoos are also doing paintings, most notably Ruby, the painting pachyderm of the Phoenix Zoo. She now uses paintbrushes filled with color. She uses her trunk to point to the brush and color she wants. If the zookeepers make a mistake and hand her the wrong brush, she simply drops the paint-laden brush to the ground to signal her displeasure.

When the calm at Ruby's zoo was disturbed by a fire truck, she painted an abstract of the truck by first choosing bright red for the color of the truck, then yellow like the emergency lights, and then blue, which was the color of the paramedics uniforms.

When Desmond Morris first showed Congo the chimp a pencil and paper, Congo played with them as he did the other objects in his environment. Then Morris wrapped the chimp's fingers around the pencil and put it against the paper and let go.

Congo brushed his hand down and then noticed something was coming out of the end of the pencil. Immediately Congo got the connection. Congo had drawn his first line.

Over the next two years, the chimp made 384 drawings and paintings. Eventually, Congo's collection was exhibited in a London art gallery.

As his involvement with art progressed, Congo, like many other artists, would throw a temper tantrum if interrupted in his work, settling down only after he was allowed to go back and complete a picture.

Penny Patterson wanted to test the color perception of Koko, the signing gorilla, so she gave Koko some paints with ten colors to choose from and asked her to paint a picture of her black-and-gray kitten who was lying on an orange towel and playing with a green toy. Koko immediately chose black for the kitten's body, then orange, and then green. Penny signed to Koko asking her about the kitten's eyes. Koko chose tan.

Humans Respond

Kindness is universal. It is a most precious part of existence and by choosing to exercise it or not, we make life heaven or hell for ourselves and our fellow beings. We have seen a few examples of the kindness demonstrated by animals for humans and other creatures. Here then, are some exceptional acts of kindness by human beings towards animals.

Sandy Mac was a hard-working horse. Daily he pulled a truck through the streets of Copenhagen. The clip-clopping of his hooves on the pavement was a familiar sound to the townsfolk in an age when automobiles were still a rarity. Sandy Mac seemed happy in his work; he had never known another life.

One day, as his driver was to go on his annual two-week vacation, the man did a rather unusual thing. He applied to give Sandy Mac some time off from the company to go on vacation with him. At first the supervisors were against the idea, but when it was pointed out to them how diligently he had worked for them for fourteen years, they agreed. And so Sandy Mac went trotting down country roads for the first time in his life. Here he encountered grass to run on and amusing chickens and geese who chased and startled him.

Finally he and his friend, the driver, arrived at the beach. Sandy Mac hated it. But when his friend dove into the water and swam away from him, the horse's good heart made him put his fears aside and rush in to save the man from what the horse thought was certain death.

When the man stood up in the water, Sandy Mac realized that the man was safe. But by that time the horse was in the water and discovered he really liked the waves. He leapt and trotted and galloped up and down the beach. When it was time to go Sandy Mac did not want to leave! For two weeks he played and chased the waves. News of Sandy Mac's vacation spread throughout Copenhagen. The local animal welfare society bought some pasture land outside the city so that other working horses could have a respite. Thereafter Sandy Mac went on vacation every year.

Edward Lear, the author of the poem "The Owl and the Pussycat," had to move from one city to another. Fearing his cat might be upset by the move, he had his new house built as an exact replica of the old one.

Despite the fact that he was left-handed, Dr. Albert Schweitzer often wrote out prescriptions with his right hand because his cat liked to sleep against his left hand and was not to be disturbed.

When the Dalai Lama won the Nobel Peace Prize in 1989, eighteen Tibetan monks celebrated by chanting to the dolphins at the Miami Seaquarium. Marine scientists said that the sound waves produced by chanting were pleasing to the dolphins who kept their heads out of the water during the performance, tilting their heads to the side to hear better.

.·' ❀ '·.

During the floods of January 1997 in Northern California, emergency helicopter rescue workers found a dog stranded on a rock. Banking the helicopter in a risky maneuver, they came down low, and a man hung far out of the aircraft and seized the dog, lifting him to safety. Many other dogs and cats were saved from rooftops during the flooding, as well.

.·' ❀ '·.

A Dutch rubber company, Dunlop-Enerka of Drachten, has started making waterbeds for cows. Initially the cows were afraid of them and did not wish to lie on them but after awhile they got to like the new beds. Working on the theory that a happy cow will produce more milk, one farmer in England ordered 180 beds. His cows liked them so much he ordered 100 more.

.·' ❀ '·.

Mohammed once carefully cut off the sleeve of his garment rather than disturb his cat, who had fallen asleep while using the ample sleeve for bedding.

·: 🐾 ·:

A study at Virginia Polytechnic Institute was conducted on chickens to see if kindness had any effect on them. Researchers spoke gently to and even sang to a group of chickens. The chickens became friendlier, put on weight faster, and were more resistant to disease than a control group that was ignored.

·: 🐾 ·:

Can you imagine massaging a rhinoceros? Elke, a massage therapist traveling in Nairobi, met a six-year-old rhinoceros named Scuddy who was injured at the knee joint. Elke had learned the "TellingtonTouch," a therapeutic technique used for both physical and behavioral treatment of animals, usually dogs, cats, and horses. The therapy involves delicate symmetrical massage motions while breathing evenly to heal and adjust the animal's subtle energy fields.

But a huge rhinoceros with its armorlike skin was an altogether new challenge. Still, it was worth a try. Elke carefully approached the injured animal and began making slow circular movements around its knee joint. She murmured reassuringly to the nervous Scuddy. When Elke came back after two days, Scuddy was very relaxed and even allowed Elke to work on her soft nostril and mouth area. By the next day of massage, Scuddy was healing so well that she could put her considerable weight on the injured leg.

Noticing that Japanese snow monkeys were shivering in the snowy winter months, villagers of northern Japan built a platform by a hot springs just for the snow monkeys. The happy monkeys were then seen playing and splashing and warming themselves in the springs during the freezing weather.

Demonstrators in England were holding their very first protest against a store that sold fur coats. When the store's owner saw the pictures they had of the painful deaths the animals endured and the gruesome conditions of even ranch-raised animals, he handed over his entire inventory of fur coats to the protesters.

Casting aside vanity and huge fees, the world's top supermodels have signed pledges not to model or sell fur coats.

.· ❀ *·.*

In California, a woman named Jane builds little homes for the animals in the forest around her home. The plastic-lined wooden boxes provide the local raccoons and skunks with shelter from the winter rains.

.· ❀ *·.*

Twenty years ago, a young orca whale was taken from his family and put in a confining tank for exhibit at a Mexican aquarium. As time passed, the young whale did not adjust well and developed a severe skin virus and other symptoms of stress. This situation might have been ignored except that this orca, Keiko, had just starred in the hit movie *Free Willy*. When news of Keiko's distress in his small tank reached the media, many forces pulled together to aid the whale, coalescing in the Free Willy Foundation. Children around the world donated a total of $100,000. Warner Brothers donated $7 million, and a new large tank that could hold saltwater was built in Oregon.

Keiko's journey from Mexico to Oregon was a heroic epic in itself. A team of US and Mexican professionals worked to ready him for the delicate transport, a long period during which he would have to survive out of water. A lift was constructed to pull him out of the water to such a height that, had he panicked and struggled, he would have been killed. United Parcel Service donated a plane and he was packed in ice. But crowds jammed the route to the airport, increasing by two hours the amount of time the whale would be out of water. Then a roller jammed and the whale could not be transferred to the plane for another two hours. Finally, he was on the plane but storms threatened to close the Oregon airport where they were to land. All these delays brought him dangerously close to death. Yet throughout his amazing ordeal, the whale remained calm and trusting.

Finally, luck was with Keiko. The skies cleared and the plane landed. The first indication of the transformation that was to come then occurred—he began to flip about in his lift with joy when he smelled the saltwater. He had never seen natural ocean water since his first capture twenty years before.

Today, for the first time in many years, Keiko can dive and really swim. His vocalizations express the great joy he feels in his new home. After six months in his new spacious tank, he has gained one thousand pounds and has grown one foot in length. His skin virus, which was extremely painful before, is almost entirely healed. A big screen TV has been installed by the tank so that he can watch and listen to videos of wild orcas. Plans are underway for the eventual full release of Keiko into the waters around Iceland from which he was captured so many years ago.

When cars first became popular, they were mostly open-air jalopies bumping along dirt roads. When one family's dog rode in the rumble seat, which he loved to do, his eyes would get full of dust. So they bought him his own pair of goggles and he rode happily along wearing his stylish goggles.

Britain's Fauna and Flora Preservation Society and Departments of Transport and Environment built that nation's first "Toad Tunnel." Noticing that about two hundred toads per hour were hopping across a certain road during their yearly migration in March, the British agencies built the tunnel so that the toads could cross safely and avoid being hit by cars.

For many years, the Board of Public Works in Los Angeles maintained a helicopter that made an "animal run" once a week to take animals that had gotten lost on city streets back to the wild.

When a horse was trapped by snowdrifts in the Colorado mountains, pilots volunteered to airlift hay to him. Many youngsters sent their chore money to pay for hay. The horse survived.

Erin Sharpe was an A student living in Flower Mound, Texas. She faced a dilemma in her young life when, while taking her high school biology course, she was told she would have to dissect a cat. Warned that her grade would suffer if she refused, she was faced with choosing between her moral standards and the grades she needed if she wanted to go to a good college.

Finally, after a long battle, the school board rescinded their decision and allowed Erin to choose an alternative. Teaching the students respect for life, it was decided, was at least as important as teaching what an animal's insides looked like.

∴   🐾   ∵

When Kim Basinger and Alec Baldwin got married
they asked that, in lieu of wedding presents,
their friends make donations to groups that
helped animals.

∴   🐾   ∵

Harry de Leyer noticed a horse that had been
passed up at a horse auction being loaded to go to
the meat packing plant. Seeing that it was facing
certain death, he bought the horse, even though
everyone thought it was a worthless, tired-out old
farm animal.

Harry brought the horse back to his ranch for
his children and his riding school. Cleaned up,
the poor old reject turned out to be white. They
named him Snow Man. Harry was amply rewarded
for his kindness. Snow Man showed promise as a
jumper, so Harry began training him. Snow Man
proved to be a natural at and went on to take the
Championship at Madison Square Garden.

∴   🐾   ∵

The Bishnoi community of India has enacted laws
to protect wildlife by forbidding hunting and the
cutting of trees. "The basic concept is to treat all
living things with dignity," they said.

In the early part of the twentieth century there were only fifteen whooping cranes left in the world. So in the 1970s, when a young crane, a female named Tex, was hatched in the San Antonio Zoo, she was reared carefully by hand.

One unexpected result of this existence was that she became accustomed to humans and refused to mate with a male whooping crane. Because female whooping cranes ovulate only when sexually excited by the male whooping crane's dance, Tex could not even be artificially inseminated.

Finally, after Tex had repeatedly turned down the overtures of a male crane, the head of the zoo decided to court her himself. It was reasoned that since she had been raised by humans, she thought herself human and therefore might respond to one. And so it was that George Archibald, a man obviously very dedicated to his job as director of the crane center, bedded down with Tex. He slept with her, talked to her, and did his best to do the wild whooping crane dance, which consists of spreading one's wings (or arms), making little pirouettes, leaping, tossing sticks in the air and catching them with one's beak (or mouth). No sluggard, Mr. Archibald also helped her collect grass for the nest.

After several failures, Tex was finally won over by these overtures and allowed herself to be artificially inseminated. She gave birth to a healthy little crane.

∴ ❀ ∴

In another breeding experiment, breeders of whooping cranes disguised themselves in sheets and hoods so they would look more like whooping cranes. They also set up cardboard crane-like cutouts so the babies would imprint on something that looked like a crane when they hatched.

∴ ❀ ∴

Honor among thieves: When robbers in New York City discovered that the van they had stolen had a dog inside, they telephoned the owners to say they had dropped the dog off at the local animal shelter.

∴ ❀ ∴

The owner of a restaurant in Rancho Cordova, California, was so moved by the protests of his customers, who had learned that a lobster he was going to serve was over one hundred years old, that he had the lobster flown back to the east coast. There it was released back into the ocean where it had lived its long life.

One year, swallows heading toward southern Europe for the winter were faced with unusually cold temperatures. As they flew over Switzerland, many couldn't make it and fell to the ground.

In an amazing concerted effort, the whole of Switzerland was mobilized into action. Schoolchildren and other volunteers picked up the birds and brought them to the local airports and railway stations by car, train, and even in cages strapped upon children's bicycles. The birds were hand-fed meat on the ends of matchsticks and flown by Swissair to their destinations. One hundred thousand birds were saved.

Resource Guide

Animal Legal Defense Fund
525 East Cotati Avenue
Cotati, CA 94931
(707) 795-2533
www.aldf.org

Doris Day Animal Foundation
8033 Sunset Boulevard, Suite 845
Los Angeles, CA 90046
www.dorisdayanimalfoundation.org

The Gorilla Foundation
P.O. Box 620530
Woodside, CA 94062
(800) ME-GO-APE
www.koko.org

Greenpeace USA
702 H Street, NW, Suite 300
Washington, DC 20001
(202) 462-1177
www.greenpeace.org/usa/

The Humane Society of the United States
1255 23rd Street NW, Suite 450
Washington, DC 20037
(202) 452-1100
www.humanesociety.org

In Defense of Animals
3010 Kerner Boulevard
San Rafael, CA 94901
(415) 448-0048
www.idausa.org

People for the Ethical Treatment of Animals
501 Front Street
Norfolk, VA 23510
(757) 622-PETA
www.peta.org

Performing Animal Welfare Society
P.O. Box 849
Galt, CA 95632
(209) 745-2606
www.pawsweb.org

Sources

The author wishes to gratefully acknowledge the wonderful books and news sources cited here. Many are delightful and highly recommended for those wishing to learn more about animals and their behavior.

Adamson, Joy. *Living Free*. New York: Harcourt, Brace & World, 1961.

Amory, Cleveland. *Animail*. New York: E.P. Dutton & Co, Inc., 1976.

---. *Man Kind?* New York: Harper and Row, 1974.

Bardens, Dennis. *Psychic Animals*. New York: Henry Holt and Company, 1987.

Berlitz, Charles. *Charles Berlitz's World of the Incredible But True*. New York: Ballantine Books, 1991.

Boone, J Allen. *Kinship With All Life*. New York: Harper & Row, 1954.

Brown, Beth. *Dogs That Work For a Living*. New York: Funk & Wagnalls, 1970.

Campbell, Neil. *Biology*. New York: Benjamin/Cummings Publishing Co, 1990.

Carson, Gerald. *Men, Beasts, and Gods: A History of Cruelty and Kindness to Animals*. New York: Charles Scribner's Sons, 1972.

Cavalieri, Paola and Peter Singer. *The Great Ape Project*. New York: Saint Martin's Press, 1994.

Darwin, Charles. *The Descent of Man and Selection in Relation to Sex*. New York: D Appleton & Co, 1896.

de Waal, Frans. *Good Natured: The Origins of Right and Wrong in Humans and Other Animals*. Cambridge: Harvard University Press, 1996.

Dickson, Lovat. *Wilderness Man*. New York: Atheneum, 1973.

Ehmann, James and David Gucwa. *To Whom it May Concern: An Investigation of the Art of Elephants*. New York: Penguin Books, 1985.

Evans, Rose. *Friends of All Creatures*. San Francisco: Sea Fog Press, Inc., 1984.

Fogle, Bruce. *Pets and Their People*. New York: Viking Press, 1984.

Frank, Morris and Blake Clarke. *First Lady of the Seeing Eye*. New York: Holt, Rinehart & Winston, 1957.

Gaddis, Vincent and Margaret. *The Strange World of Animals and Pets*. Stamford, CT: Cowles Book Company, Inc., 1970.

Gellius, Aulus. *The Attic Nights of Aulus Gellius*. Cambridge: Harvard University Press, 1952.

George, Dick. *Ruby: The Painting Pachyderm of the Phoenix Zoo*. New York: Delacorte Press, 1995.

Gonzalez, Philip and Leonore Fleischer. *The Dog Who Rescues Cats*. New York: HarperCollins, 1995.

Goodall, Jane. *With Love*. Jane Goodall Institute, 1994.

Greene, Lorne. *The Lorne Greene Book of Remarkable Animals*. New York: Simon & Schuster, 1980.

Helfer, Ralph. *The Beauty of the Beasts*. Los Angeles: Jeremy P. Tarcher, 1990.

Hyde, Dayton. *Sandy: The True Story of a Rare Sandhill Crane*. New York: Dial Press, 1968.

Kowalski, Gary A. *The Souls of Animals*. Walpole, NH: Stillpoint Publishing, 1991.

Masson, Jeffrey Moussaieff and Susan McCarthy. *When Elephants Weep*. Delacorte Press, 1995.

McElroy, Susan Chernak. *Animals As Teachers and Healers*. NewYork: Sage Press, 1996.

Montgomery, Sy. *Walking with the Great Apes*. New York: Houghton Mifflin, 1991.

Morris, Desmond. *Animal Days*. New York: William Morrow and Company, 1980.

Newkirk, Ingrid. *Kids Can Save the Animals*. New York: Warner Books, 1991.

---. *Save the Animals*. New York: Warner Books, 1990.

Patterson, Francine. *Koko's Kitten*. New York: Scholastic, Inc., 1985.

Rasa, Anne. *Mongoose Watch*. New York: Anchor Press, 1986.

Reader's Digest Association, Inc. *Animals You Will Never Forget*. 1969.

Reader's Digest Association, Inc, ed. *Animals Can Be Almost Human*. 1979.

Regenstein, Lewis. *The Politics of Extinction*. New York: MacMillan Publishing, 1975.

Ring, Elizabeth. *Assistance Dogs in Special Service*. Brookfield, CT: The Millbrook Press, 1993.

Robbins, John. *Diet for a New America*. Walpole, NH: Stillpoint Publishing, 1987.

Roberts, Yvonne. *Animal Heroes*. London, UK: Pelham Books, 1990.

Romanes, George John. *Animal Intelligence*. Washington DC: University Publications of America, 1977.

Ruesch, Hans. *Slaughter of the Innocent*. New York: Bantam Books, 1978.

Singer, Peter. *Animal Liberation*. New York: Avon Books, 1975.

Stewart, John. *Elephant School*. New York: Pantheon Books, 1982.

Terres, John K. *The Audobon Book of True Nature Stories*. Thomas Y. Crowell Co., 1958.

Thomas, Elizabeth Marshall. *The Hidden Life of Dogs*. New York: Houghton Mifflin Co, 1993.

Wallace, Irving, David Wallechinsky, Amy Wallace, and Sylvia Wallace. *The People's Almanac Presents the Book of Lists # 2*. New York: Morrow, 1980.

---. *The People's Almanac Presents the Book of Lists: The 90's Edition*. Boston: Little, Brown & Company, 1993.

---. *The People's Almanac Presents the Book of Lists*, New York: Morrow, 1977.

Watson, E.L.Grant. *Animals in Splendour*. Camp Hill, PA:
Horizon Press, 1967.

Wels, Byron G. *Animal Heroes: Stories of Courageous
Family Pets and Animals in the Wild*. New York:
Macmillan, 1979.

White, Betty and Thomas J. Watson. *Betty White's Pet-
Love*. New York: William Morrow and Company, 1983.

Williams, Heathcote. *Sacred Elephant*. New York:
Harmony Books, 1989.

Wood, Gerald L. *Guinness Book of Pet Records*. London,
UK: Guinness Superlatives Ltd, 1984.

Wylder, Joseph. *Psychic Pets Secret Life*. New York:
Random House, 1978.

Yerkes, Robert M. and Ada W. *The Great Apes: A Study
of Anthropoid Life*. New Haven: Yale University
Press, 1929.

Additional Sources

BBC Wildlife

The Journal of Mammology by David H. Brown and
   Kenneth S. Norris

Catnip Magazine

CNN

Country Magazine

Elke Riesterer

In Defense of Animals: The IDA Magazine

Ken-L-Ration Dog Hero of the Year Awards

Life Magazine

National Geographic

People Magazine

PETA's Animal Times Magazine

Sharon Callahan and *Mt Shasta Directions*

The Free Willy Story: Keiko's Journey Home

Mango Publishing, established in 2014, publishes an eclectic list of books by diverse authors—both new and established voices—on topics ranging from business, personal growth, women's empowerment, LGBTQ studies, health, and spirituality to history, popular culture, time management, decluttering, lifestyle, mental wellness, aging, and sustainable living. We were recently named 2019 and 2020's #1 fastest-growing independent publisher by *Publishers Weekly*. Our success is driven by our main goal, which is to publish high-quality books that will entertain readers as well as make a positive difference in their lives.

Our readers are our most important resource; we value your input, suggestions, and ideas. We'd love to hear from you—after all, we are publishing books for you!

Please stay in touch with us and follow us at:
    Facebook: Mango Publishing
    Twitter: @MangoPublishing
    Instagram: @MangoPublishing
    LinkedIn: Mango Publishing
    Pinterest: Mango Publishing
    Newsletter: mangopublishinggroup.com/newsletter

Join us on Mango's journey to reinvent publishing, one book at a time.